Wastewater Microbiology

A Handbook for Operators

Toni Glymph

American Water Works Association

Science and Technology

AWWA unites the drinking water community by developing and distributing authoritative scientific and technological knowledge. Through its members, AWWA develops industry standards for products and processes that advance public health and safety. AWWA also provides quality improvement programs for water and wastewater utilities.

Printed in the United States of America.

Disclaimer

The authors, contributors, editors, and publisher do not assume responsibility for the validity of the content or any consequences of their use. In no event will AWWA be liable for direct, indirect, special, incidental, or consequential damages arising out of the use of information presented in this book. In particular, AWWA will not be responsible for any costs, including, but not limited to, those incurred as a result of lost revenue. In no event shall AWWA's liability exceed the amount paid for the purchase of this book.

Library of Congress Cataloging-in-Publication Data has been applied for.

American Water Works Association

6666 West Quincy Avenue
Denver, CO 80235-3098
303.794.7711

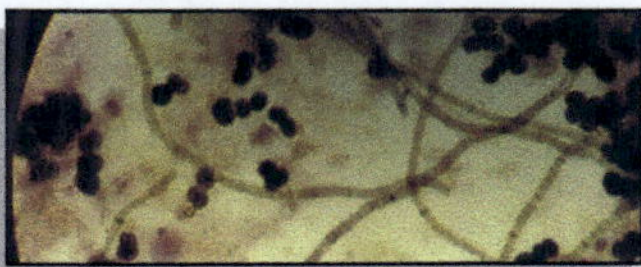

Contents

Preface

From the moment I realized that wastes from homes did not simply disappear into the ground and began to understand that wastewater treatment is a fascinating and very complicated scientific process, my curiosity was peaked and my interest stirred. Wastewater treatment presented yet another very significant contribution that microorganisms make in our society. My love for microbiology led me to study the role of microorganisms in wastewater treatment. I discovered their significance and learned to appreciate the role they play. I began to understand the conditions under which they do their best work as well as the conditions that hinder their performance. I also realized that their role was not widely understood or appreciated.

This book is written to highlight the value of the microorganisms in the wastewater treatment process and to assist the wastewater operator in understanding the role of microbiology in process control. It is my hope that as you read this book you will gain an appreciation for the rich microlife in your treatment plant and that your curiosity will be peaked and your interest stirred.

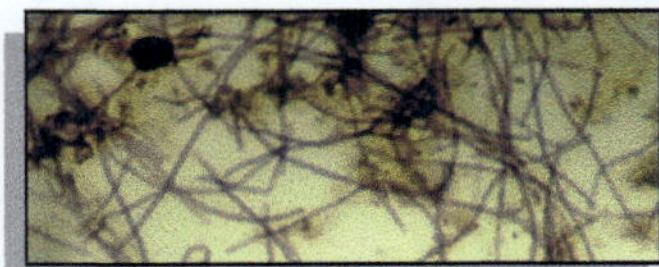

Acknowledgments

To God goes all the glory.

First, I thank my husband, Rubin: You are the wind beneath my wings. I thank my sons Lawrence and Daniel, who were patient during the months I sat in front of the computer. Thank you Cheryl; you were a vessel of mercy in a dry place. Thank you Jon and Keith. This is as much your book as mine. And, thank you Delilah for your administrative support and Jack for your expert review. Thank you Kari for being my friend and colleague. Thank you Theola. Your enthusiasm really encouraged me. Thank you Andre Jackson for your wonderful illustrations. I give special thanks to Curtis Truss, director of the Operator Training Committee of Ohio, and the wastewater treatment plant professionals, especially those at the Alum Creek Wastewater Treatment Plant, who volunteered their sludge and encouraged me to write this book. This is my gift to you! To all, thanks and God bless you. I could not have done it without you.

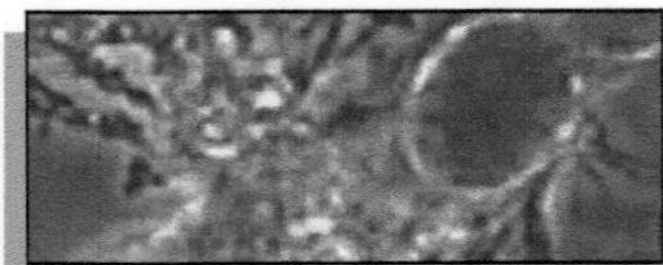

Introduction

Why microbiology? Because wastewater treatment is a biological process. A wastewater treatment plant is a microbiological zoo, so to speak, that houses bacteria, protozoa, metazoa, and other microlife. The microorganisms break down and remove nutrients and organic material in the wastewater. They perform best when all their needs are met, just like you and me.

The operator's task is to provide a favorable environment for the microorganisms. It is actually a win–win partnership. The operator provides the food and the favorable environment, and the microorganisms remove nutrients and provide wastewater treatment.

Wastewater is generated from many sources. Volumes of wastewater, called *sewage*, are generated each time it rains, or when we flush a toilet, take a shower, or wash dishes—anytime water is used. You can even predict commercial breaks during a Super Bowl game; the sewage flow increases dramatically when everyone rushes to the bathroom during a commercial. Each day, the amount of wastewater generated can range from thousands of gallons in a small community to billions of gallons in a large city. This sewage represents thousands of tons of organic matter. Where does all this water go? Allowing this untreated discharge (which is rich in organic nutrients and microorganisms) to enter

nearby lakes and streams would cause an overgrowth of algae and other aquatic vegetation and a subsequent loss of oxygen that would result in massive fish kills. Instead, sewage is routed through wastewater treatment facilities where undesirable materials and harmful microorganisms are removed.

Wastewater treatment facilities are designed to allow pollution to break down naturally under controlled conditions. These systems include physical and chemical processes to remove solids and heavier materials. However, left behind is the liquid that contains soluble and insoluble organic materials. The one process all sewage facilities have in common is the biological treatment of this organic material. That is, sewage facilities rely on certain microorganisms to convert these organic nutrients into beneficial materials. Wastewater treatment systems are designed to house these microorganisms and bring them in contact with the organic nutrients in sewage. Sewage contains nutrients of every type: phosphorus, nitrogen, sodium, potassium, iron, calcium, and compounds such as fats, sugars, and proteins. Microorganisms use these substances as food sources for energy, to synthesize cell components, and to maintain life processes. The health and well-being of these microorganisms are critical to the adequate treatment of sewage.

As environmental regulations become stricter and eliminate many currently used chemical treatment processes, wastewater treatment plant operators must be knowledgeable and trained to maintain an environment within the wastewater treatment system that will encourage the growth of the microorganisms best suited to treat the water.

Biological treatment of phosphorus and ammonia is one of the current trends toward eliminating the need for chemical treatment. There is no question that biological treatment is better for the environment. To stay ahead of regulations, it is important for wastewater treatment professionals to understand how the microorganisms function and the environmental factors that are required to enable them to best treat the wastewater.

Chapter 1

Wastewater Treatment System Overview

Humans can accumulate a mass of wastes. Clothes, remains of food, corpses, pet wastes, hair, fingernails—the list goes on. Without bacteria and other microbes, we would be knee deep in our own waste products. These microorganisms play a primary role in transforming wastes into harmless or reusable materials. The most important example of this valuable function is wastewater treatment.

Earlier civilizations, especially during the Middle Ages, used steps, corners, bushes, or any odd corner as lavatories. Wastes were thrown into the streets and disease became rampant. As populations increased, so did the need to deal with waste. By the early twentieth century sewage was being collected and piped to farms and nearby creeks and used as fertilizer. This sometimes contaminated drinking water. Sewage techniques have made enormous strides; they tap into natural biological processes and substantially reduce the risk of contamination.

Sewage is an extremely satisfactory medium for the growth of microorganisms. Wastewater treatment systems are designed to take advantage of the natural biological processes by providing this medium in an environment that will encourage their growth.

Preliminary Treatment

Municipal wastewater can be a mixture of domestic and industrial wastes. These wastes are discharged into the sanitary sewer. Storm water that is not discharged through a separate storm sewer is sometimes mixed with domestic waste. The combination of domestic wastewater and storm water is discharged through a combined sewer.

Influent wastewater contains large debris. Rags, rocks, cans, sticks, branches, and leaves, along with grit, sand, dirt, and other types of garbage, enter the treatment system through interceptors. In spite of all this debris, wastewater is still 99.9 percent water. During preliminary treatment, the wastewater flows through screens and bars where the larger debris is trapped. These larger materials that would otherwise interfere with pumps and clog the flow of wastewater through the system are removed.

The aerated grit removal system is another part of preliminary treatment. Here, the wastewater stream is slowed and slightly agitated (usually with air) to allow grit to settle. At the same time, most of the organic materials are kept in suspension. Without this removal process, grit would otherwise soon settle out and fill downstream tanks, rendering them unusable.

Primary Treatment

Primary treatment is designed to "float out the floatables and settle out the settleables." In the early twentieth century, this was the extent of wastewater treatment. During primary treatment, wastewater flow is slowed even more and held in primary settling tanks. Here greases and oils float to the top and are skimmed off for disposal. Heavier solid materials settle to the bottom. Polymer is added to some primary systems to help settle suspended solids. About two thirds of the suspended solids, oils, and greases and about one third of the organic materials are removed during primary treatment (primary effluent), resulting in an effluent that contains mostly dissolved organic wastes.

Secondary Treatment

Secondary treatment is a biological process designed to promote the growth of microorganisms that digest dissolved and finely suspended materials. Secondary treatment, along with primary treatment, can reduce the suspended solids and nutrient loads by 85 percent. Modifications to this system can also reduce ammonia, nitrates, and phosphorus.

There are two distinct types of secondary treatment systems: fixed film and activated sludge. Fixed film treatment systems use microorganisms that grow on the surfaces of rocks or plastic to treat the water. In the activated sludge treatment system, primary effluent enters the secondary aeration basin where it is mixed with microorganisms, whose primary role is to remove most of the remaining dissolved materials. This mixture of microorganisms and wastewater is called *mixed liquor*. The mixed liquor then enters settling basins or *clarifiers* where the biological solids (most often called *sludge* or *biosolids*) settle out. Some of this microorganism-rich sludge is returned to the head of the aeration basin and is mixed with incoming wastewater. The remaining portion of the sludge is wasted out to the solids handling process. The remaining liquid, called *secondary effluent*, is relatively free of dissolved organic materials. It is then disinfected or simply discharged into a receiving water body (Figure 1-1).

The activated sludge treatment process can be set up in different ways. For instance, in the plug-flow arrangement the wastewater and the returned microorganisms (sludge) are added at the beginning of the aeration basin. The wastewater and return sludge flow together through three or four long passes before being discharged into the clarifier (Figure 1-2). In the step-feed arrangement, wastewater or return sludge is added to the aeration basin in more than one step or location along the basin (Figure 1-3). In the contact stabilization process, the return sludge is routed to a stabilization or re-aeration tank and is aerated before being added to the wastewater (Figure 1-4).

These and other arrangements, such as tapered aeration, oxidation ditches, sequencing batch reactors, aerated lagoons, and

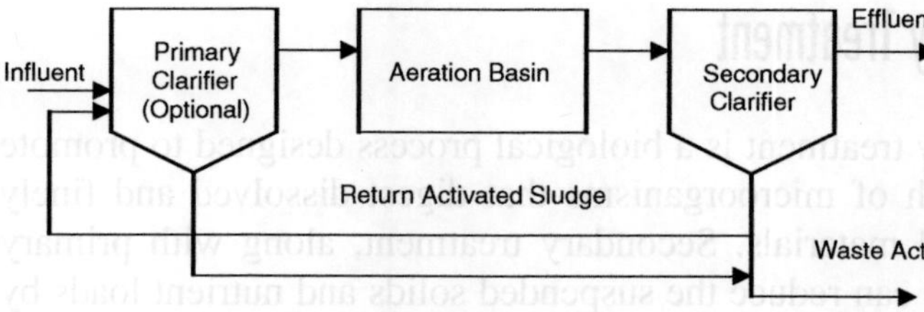

Figure 1-1 Conventional activated sludge design.

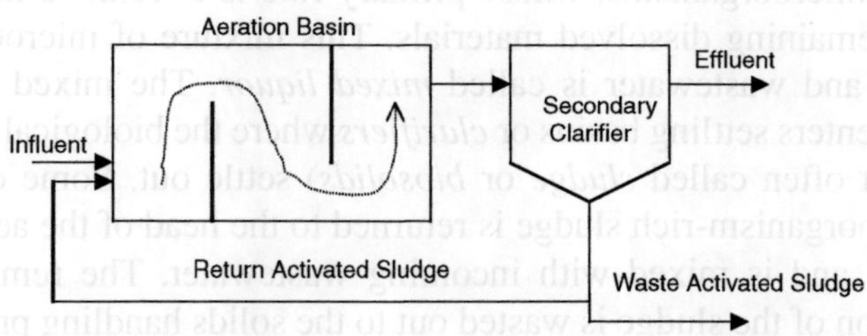

Figure 1-2 Plug-flow activated sludge design.

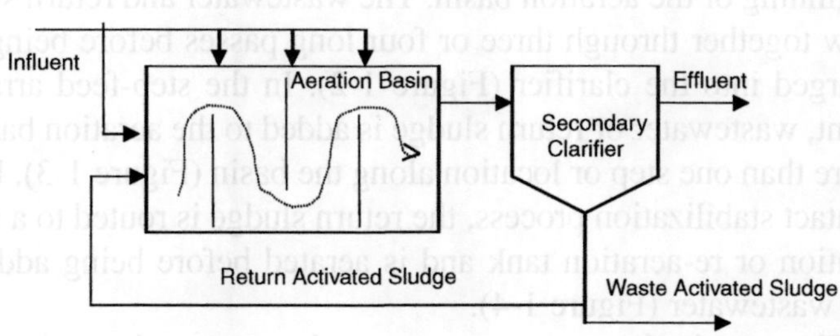

Figure 1-3 Step-feed activated sludge design.

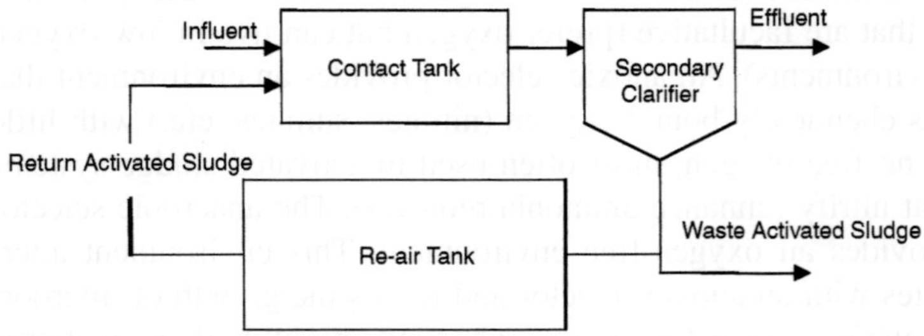

Figure 1-4 Contact stabilization activated sludge design.

extended aeration, are all variations of the activated sludge treatment process. Each has advantages and disadvantages. However, one thing is common to all process variations: microorganisms are brought in contact with the wastewater, which serves as a food source. The microorganisms remove the organic materials from the wastewater and settle out as sludge.

Wastewater treatment systems are designed to provide an environment that encourages the growth of desirable microorganisms and discourage the growth of undesirable ones. In other words, the operator is purposely selecting for microorganisms that are better suited to provide the desired treatment. Although in the conventional activated sludge treatment system, some selection of microorganisms naturally occurs, advances in wastewater treatment science allow the operator to select microorganisms even more specifically.

Selectors

A selector is an area of the treatment system that is designed to select for a specific type of microorganism by providing an environment that is favorable for its growth. A selector can be either a separate compartment in the aeration basin or a single tank located in front of the aeration basin. Selectors are designed to create an environment that is different from that in the aeration basin.

The selector environment can be aerobic, much like the aeration basin that provides free or chemically bound oxygen. This

environment selects for microorganisms that are strictly aerobic or that are facultative (prefer oxygen but can live in low-oxygen environments). An anoxic selector provides an environment that has chemically bound oxygen (nitrates, sulfates, etc.) with little or no free oxygen, most often used in activated sludge systems that nitrify (enhance ammonia removal). The anaerobic selector provides an oxygen-free environment. This environment alternates with an anoxic selector and favors the growth of microorganisms connected with enhanced biological phosphorus removal. Selectors have also been used to discourage the growth of filamentous bacteria and may be used to reduce the total nitrogen in the wastewater.

Whether you use selectors or a conventional activated sludge system, knowledge and understanding of how the microorganisms function under different conditions are key to the optimal treatment of wastewater.

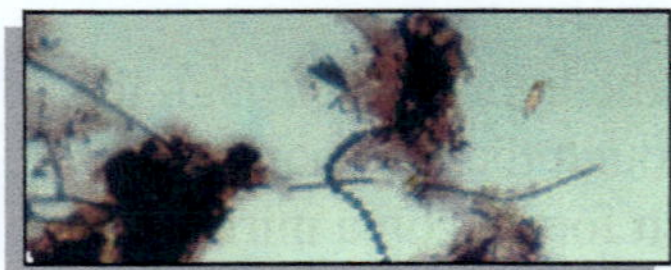

Chapter 2

General Microscopy

Microorganisms are too small to see with the naked eye, which can only see things as small as 0.1 mm. Most bacteria are 0.001 mm. If you take 1,000 bacteria cells and lay them end-to-end, they would only span the tip of a pin (Postgate 2000). Therefore, their activity can be viewed only through a microscope.

The Microscope

In the late 1600s, Anton Van Leeuwenhoek was one of the first to observe what he called *animalcules* through a simple, single-lens microscope, very much like a magnifying glass. Today microscopes can magnify from 10 to thousands of times more than Van Leeuwenhoek's single-lens microscope. In 1830 Joseph Jackson Lister designed a microscope that used a combination of lenses. This led to the compound microscope used today. A compound microscope has an ocular lens and an objective lens. The ocular lens is the eyepiece. Some microscopes are *mon*ocular, with one eyepiece; some are *bin*ocular, with two eyepieces; and some are *tri*nocular, with three eyepieces. Trinocular microscopes allow a second person to view the specimen or a camera to be attached to photograph the specimen.

A compound microscope allows tiny microorganisms to be examined in detail. Light from an illuminator, or light source, is

passed through a condenser that directs the light rays through the microorganisms. The light rays pass through the microorganisms into the objective lens. The image is then formed on a mirror and is magnified again by the ocular lens (Figure 2-1).

Most ocular lenses magnify 10×. There are usually three objective lenses: 10× (low power), 40× (high power), and 100× (oil immersion). The oil immersion lens is used by lowering the objective and immersing it in oil before focusing it on the specimen.

How is the total magnification of a specimen determined? If you multiply the magnification of an objective with the magnification of the ocular, you can determine the total magnification of the specimen. For example, for the low-power objective, the total magnification would be 10 (ocular magnification) times 10 (low-power objective) and would equal 100. For the high-power objective, the total magnification would be 400; for the oil immersion objective, 1,000.

Under normal operation, light rays from a light source are passed through a *condenser* that directs the rays through the specimen. This produces a *brightfield* illumination. However, most microorganisms in wastewater are transparent. This makes distinguishing cell structures difficult, because the water is also

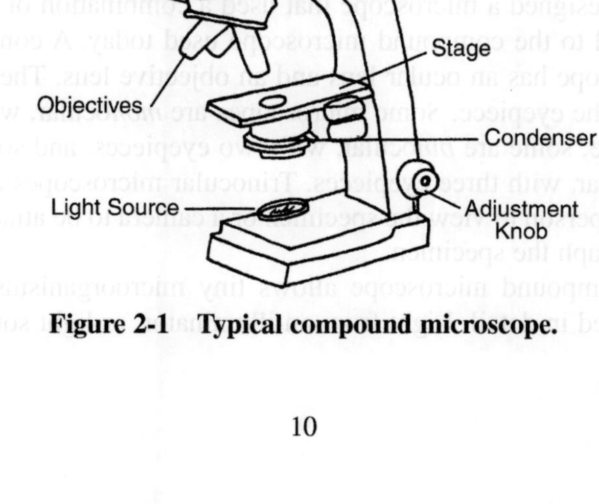

Figure 2-1 Typical compound microscope.

transparent. Several microscopy techniques allow clear and distinct cell structures to be seen.

Phase contrast microscopy uses a special condenser that slows the light as it enters the denser parts of the microorganisms. This allows certain structures to stand out from other less dense parts of the cell and from the surrounding fluid. This technique allows living organisms to be observed. Cell shape and structure are much more visible than with brightfield illumination (Figure 2-2).

Another technique, *darkfield microscopy*, makes the microorganisms appear white and the background black. The light does not pass through the specimen, but hits only its sides (Figure 2-3).

Differential interference contrast (DIC) microscopy gives a three-dimensional appearance of the cell and other cell structures. A DIC microscope contains specially designed prisms that recombine split beams of light to give a three-dimensional effect (Figure 2-4).

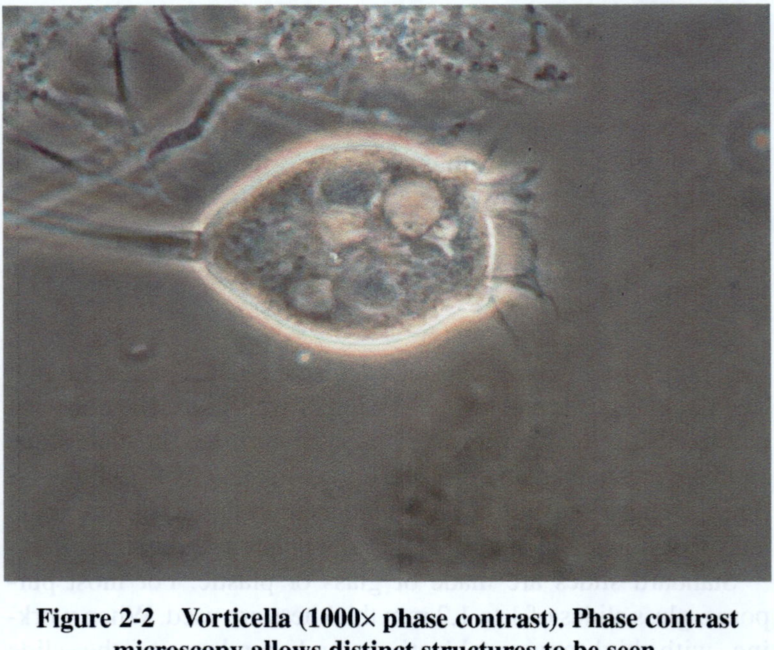

Figure 2-2 Vorticella (1000× phase contrast). Phase contrast microscopy allows distinct structures to be seen.

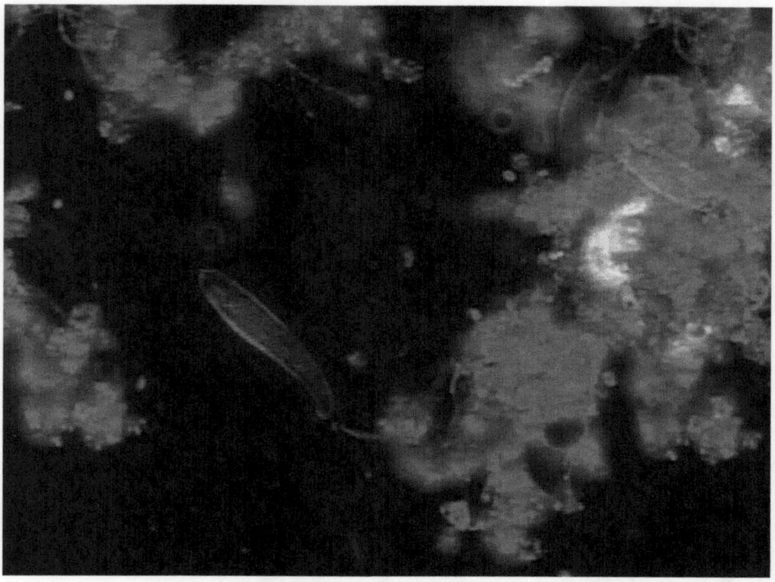

Figure 2-3 Darkfield microscopy. The organisms appear white; the background appears black.

Electron microscopy is used to examine much smaller objects such as viruses or the internal structures of cells. The electron microscope uses beams of electrons instead of light. The magnifying power of the electron microscope is far greater than that of all other types of microscopes.

Preparing the Slide

There are two different types of microscope slides in general use. The flat glass slide and the depression or well slide. The one most commonly used for wastewater observations is the flat glass slide. Both are rectangular and measure approximately 25 × 75 mm (1 × 3 in.). Depression slides have an indentation in the center to hold a drop of liquid and are usually used without a cover slip.

Standard slides are made of glass or plastic. For most purposes, glass slides of 1 to 1.2 mm thickness are used. When working with high-power objectives and condensers, the slide

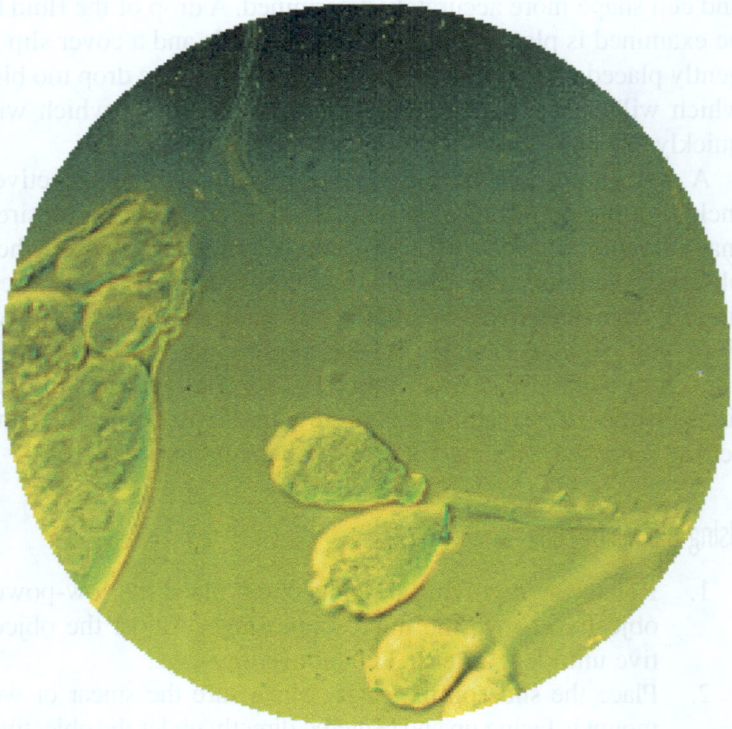

Figure 2-4 Differential interference contrast (DIC) microscopy produces a three-dimensional image.

thickness should be reduced to 0.8 to 1 mm. A cover slip or cover glass is a very thin square piece of glass (or plastic) that is placed over the water drop on the slide. With a cover slip in place, the drop is flattened out, allowing the observer to focus with high power on the specimen. The cover slip also confines the specimen to a single plane, thereby reducing the amount of focusing necessary. The cover slip also protects the objective lens from immersion into the water drop.

Wet Mount

A wet mount allows living microorganisms to be observed as they appear in the environment. Cell measurements can be taken

and cell shape more accurately determined. A drop of the fluid to be examined is placed in the middle of a slide and a cover slip is gently placed on top. It is important not to make the drop too big, which will make the cover slip float, or too small, which will quickly dry up (Figure 2-5).

A wet mount can be observed using any of the objectives including the 100× objective with oil depending on the desired magnification. It is best to use the phase contrast condenser when observing samples live. The phase contrast condenser has a setting for each objective. Each time you switch to a different objective, you should change to the appropriate phase contrast condenser setting. Usually, setting 1 is for the lower magnifications (10× or 20×), setting 2 is used with the 40× objective, and setting 3 is used with the 100× oil immersion lens objective.

Using the Microscope

1. Rotate the revolving nosepiece and place the low-power objective above the microscope stage. Adjust the objective until it is at least 3 cm above the stage.
2. Place the slide on the stage. Make sure the smear or wet mount is facing up and centered directly under the objective.
3. Peering through the eyepiece, turn the coarse adjustment knob and move the objective toward the slide until the specimen begins to come into focus.
4. Turn the fine adjustment knob to bring the specimen into sharp focus.
5. Once the specimen is focused with the low-power objective, rotate the nosepiece to place the higher-power objective over the specimen. The specimen should be

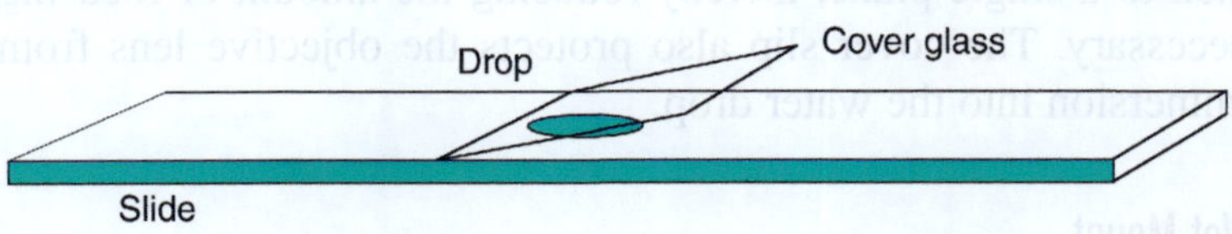

Figure 2-5 Wet mount. A drop of mixed liquor is placed on a slide and is covered with a cover slip. This technique is used to observe live specimens, cell shapes, and structures.

nearly in focus. Adjust the focus with the fine adjustment knob.

6. To move to the oil immersion lens, rotate the nosepiece until it is between objectives. Place a drop of oil directly on the stained smear or the cover slip on the wet mount. Slowly rotate the 100× objective into the oil. The image should be nearly in focus. Bring the image into sharp focus with the fine adjustment knob.

Smear Preparation and Staining

Microscopic cells, tissues, and organisms are transparent and can be difficult to see. Important information such as cell shape and motility can be observed without stains, but staining reveals different properties and provides contrast for viewing. A stain is a substance that adheres to or penetrates a cell to give it color. Different stains are attracted to different types of organisms or different parts of an organism and can be used to differentiate between organisms or to view specific parts of organisms.

To stain a cell, fix the microorganisms to the slide so that they will not wash off with water. This is done by spreading a thin film of mixed liquor or foam over the slide and allowing it to air dry. Once the sample has dried completely, it is ready for staining (Figure 2-6).

The two most common staining techniques are simple and differential. Simple staining uses only one color and does not differentiate between microorganisms. Differential staining uses more than one dye and will stain different types of microorganisms different colors. The bacteria types are distinguished by the way they react to the stains. Differential staining is used more often in wastewater microbiology. The most common differential stain used is the Gram stain.

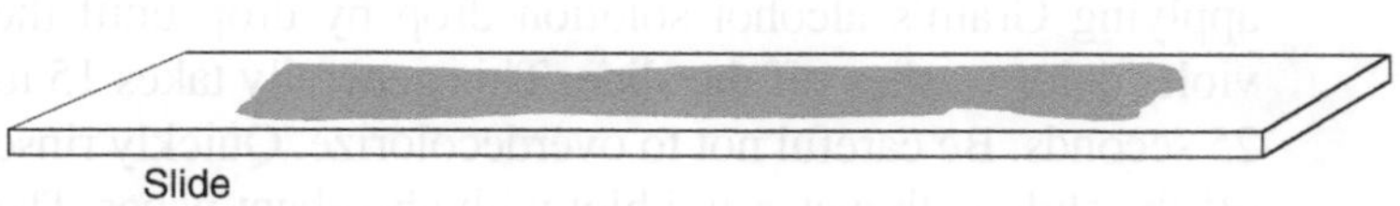

Figure 2-6 Smear. A drop of sample is gently smeared across the slide and allowed to air dry before staining.

Gram Stain

The Gram staining procedure was developed in 1884 by Hans Christian Gram. It is probably the most widely used procedure and helps to distinguish between two types of bacteria that otherwise would be difficult to tell apart. It divides bacteria into two large groups: (1) gram-positive bacteria that retain the color of the crystal violet dye and (2) gram-negative bacteria that lose the crystal violet color when washed with alcohol. The gram-negative microorganisms are then counterstained with safranin, which gives them a pink color. Thus, gram-positive bacteria stain purple and gram-negative bacteria stain pink. Bacteria react differently to staining because of the difference in the structures of their cell walls. The cell wall on a gram-negative bacterium has a layer made up of lipopolysaccharide. This polymer-like substance resists the purple dye so the cell wall will not retain the purple color.

The Gram staining procedure is done in four simple steps with four staining solutions, which can be purchased already prepared. Make sure you are staining the right side of the slide (smear side up)!

Procedure

1. Completely flood the smear with crystal violet and let it stand for 1 minute. Gently wash the slide with water. Gram-negative and gram-positive bacteria become directly stained and appear purple (Figure 2-7).
2. Completely flood the smear with Gram's iodine solution and let it stand for 1 minute. Gently wash the slide with water and blot with absorbent paper. A crystal violet-iodine complex is formed that helps the bacteria retain the purple color.
3. Decolorize the smear by holding the slide at an angle and applying Gram's alcohol solution drop by drop until the violet color washes off the slide. This generally takes 15 to 25 seconds. Be careful not to overdecolorize. Quickly rinse off the slide with water and blot with absorbent paper. The gram-positive cells remain purple and the gram-negative cells become colorless (Figure 2-8).

Figure 2-7 Gram stain step 1. The smear is covered with crystal violet. Gram-positive and gram-negative bacteria become directly stained and appear purple.

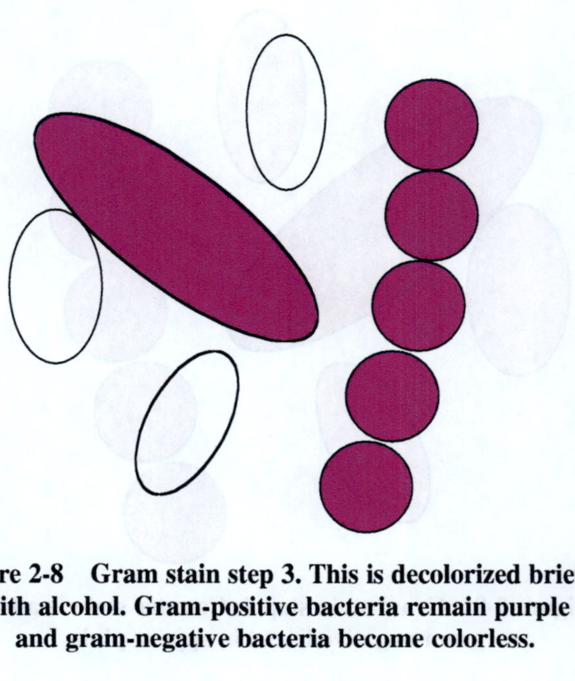

Figure 2-8 Gram stain step 3. This is decolorized briefly with alcohol. Gram-positive bacteria remain purple and gram-negative bacteria become colorless.

4. Flood the smear with the counterstain safranin and let it stand for 1 minute. Rinse well with water and blot dry. The colorless gram-negative bacteria are stained pink by the safranin (Figure 2-9).
5. Examine under oil immersion using the 100× objective with brightfield illumination (do not use phase contrast).

Neisser Stain

Another differential stain that is often used with activated sludge samples is the Neisser stain. This stain was named after Albert Neisser, "affectionately" called the Father of Gonococcus, the bacterium responsible for gonorrhea. He was also the first to stain and demonstrate the leprosy bacterium. The Neisser stain also divides bacteria into two large groups, those that store polyphosphate granules and those that do not. Neisser-positive bacteria stain blue; Neisser-negative bacteria stain brown. Neisser-positive granules can be seen within the cells of some bacteria.

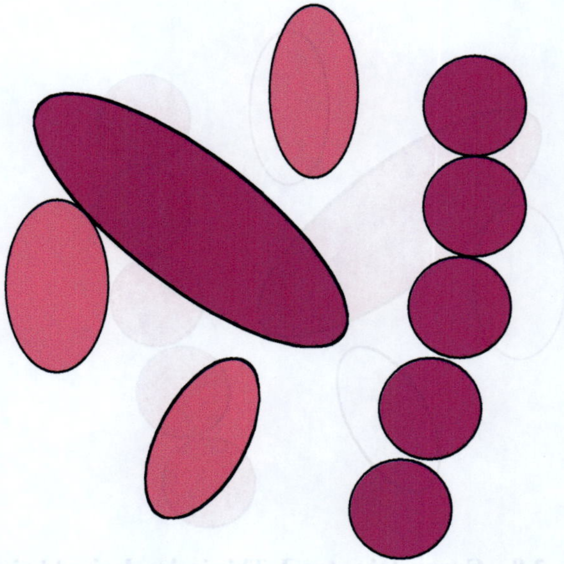

Figure 2-9 Gram stain step 4. Counter stain is added; gram-negative bacteria appear pink.

Neisser staining solutions can also be purchased in a kit already prepared. The kit contains three solutions: methylene blue, crystal violet, and Bismark brown.

Procedure

1. Prepare solution 1 by mixing two parts by volume of methylene blue with one part by volume of crystal violet. This solution must be prepared monthly.
2. Cover a smear that has been thoroughly air dried with solution 1 for 30 seconds and rinse with water for 1 to 2 seconds.
3. Cover the smear with Bismark brown for 1 minute, rinse well, and blot dry.
4. Examine with 100× oil immersion with brightfield illumination (do not use phase contrast).

References

Postgate, J. 2000. *Microbes and Man*. Cambridge, Mass.: Cambridge University Press.

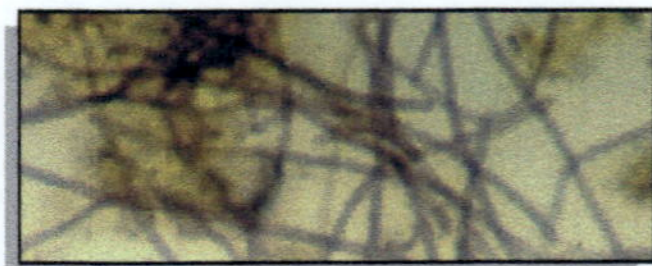

Chapter 3

Wastewater Microorganisms

In the activated sludge process, microorganisms are mixed with wastewater. The microorganisms come in contact with the biodegradable materials in the wastewater and consume them as food. In addition, the bacteria develop a sticky layer of slime around the cell wall that enables them to clump together to form biosolids or sludge that is then separated from the liquid phase. The successful removal of wastes from the water depends on how efficiently the bacteria consume the organic material and on the ability of the bacteria to stick together, form floc, and settle out of the bulk fluid. The flocculation (clumping) characteristics of the microorganisms in activated sludge enable them to amass to form solid masses large enough to settle to the bottom of the settling basin. The better the flocculation characteristics of the sludge, the better the settling and the better the wastewater treatment.

After the aeration basin, the mixture of microorganisms and wastewater (mixed liquor) flows into a settling basin or clarifier where the sludge is allowed to settle. Some of the sludge volume is continuously recirculated from the clarifier back to the aeration basin to ensure adequate amounts of microorganisms are maintained in the tank. The microorganisms are again mixed with incoming wastewater where they are reactivated to consume organic nutrients. Then the process starts again.

The activated sludge process, under proper conditions, is very efficient. It removes 85 to 95 percent of the solids and reduces the biochemical oxygen demand (BOD) about the same amount. The efficiency of this system depends on many factors, including wastewater climate and characteristics. Toxic wastes that enter the treatment system can disrupt the biological activity. Wastes heavy in soaps or detergents can cause excessive frothing and thereby create aesthetic or nuisance problems. In areas where industrial and sanitary wastes are combined, industrial wastewater must often be pretreated to remove the toxic chemical components before it is discharged into the activated sludge treatment process. Nevertheless, microbiological treatment of wastewater is by far the most natural and effective process for removing wastes from water.

Activated Sludge Microorganisms

Five major groups of microorganisms are generally found in the aeration basin of the activated sludge process:

1. Bacteria
2. Protozoa
3. Metazoa
4. Filamentous bacteria
5. Algae and fungi

Bacteria make up about 95 percent of all the microorganisms in activated sludge. They have the primary role in removing the nutrients from the wastewater. Let's study them in more detail in the next five chapters.

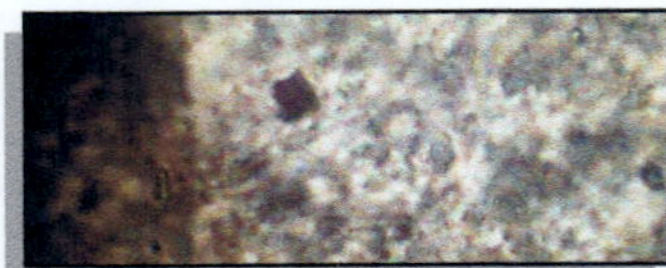

Chapter 4

Bacteria

Bacteria are amazingly complex, even though each has only a single cell. Some bacteria can grow in extreme cold or extreme heat—temperatures beyond which any human can survive. They are among the most abundant organisms on Earth and can be found on our skin, in our mouths, in our intestines, in our food and drinks, in the air, in water, and in soil. They can be found in the highest altitudes, deepest oceans, in the Arctic, and in volcanoes.

We often associate bacteria with disease, but most are harmless and many provide valuable services for humans. Bacteria in the soil, water, and sewage systems transform human wastes into reusable materials.

Bacteria can be classified in many ways. They are frequently classified based on how they respond to oxygen. *Aerobic* bacteria require oxygen for growth and maintenance. *Anaerobic* bacteria, however, cannot tolerate oxygen. *Facultative* bacteria prefer oxygen but can survive without it.

The most important microorganisms in the activated sludge process are the aerobic bacteria. Bacteria make up about 95 percent of the microorganisms in activated sludge. They are introduced into the treatment system via surface runoff, wash water, and feces. Since they are primarily responsible for removing organic nutrients from the wastewater, it is important to understand the conditions that make them work their best. Let's begin by understanding how they are structured.

Structure

Bacteria are small, relatively simple, one-celled microorganisms. They generally have one of three basic shapes: *coccus* (spherical or oval shaped), *bacillus* (rectangular, rod-like, or discoid), and *spirillum* (spiral or corkscrew shaped). Some are individual cells; others form clusters, pairs, chains, squares, or other groupings.

Coccus-shaped bacteria (the word *coccus* means berries) occur singly or in pairs called *diplococci*; attached in short chains called *streptococci*; in groups of four called *tetrads*; in three-dimensional cubes of eight called *sarcinae*; and in grapelike clusters called *staphlycocci*. They can be both gram-positive and gram-negative (Figure 4-1).

Most bacillus-shaped bacteria (the word *bacillus* means little staff) appear as single rods. However, they sometimes occur in pairs called *diplobacilli* and short chains called *streptobacilli*. They also occur in short, almost oval rods called *coccobacilli* (Figure 4-2).

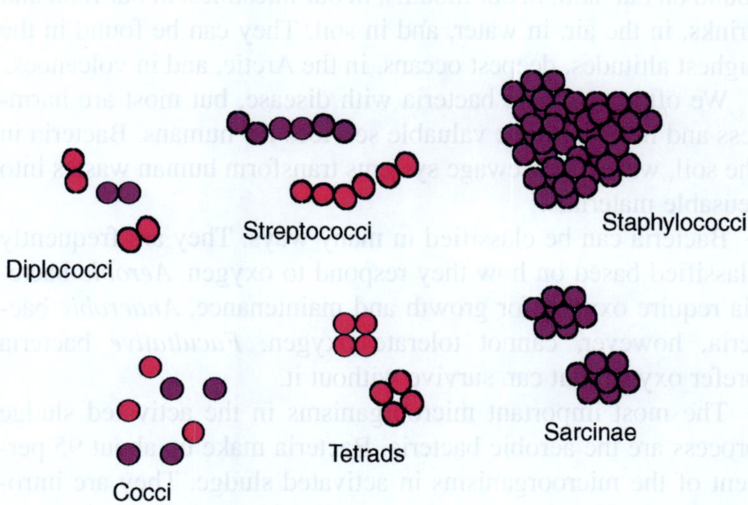

Figure 4-1 Coccus-shaped bacteria can occur in pairs called *diplococci,* short chains called *streptococci,* groups of four called *tetrads,* cubes of eight called *sarcinae,* and grape-like clusters called *staphylocci.*

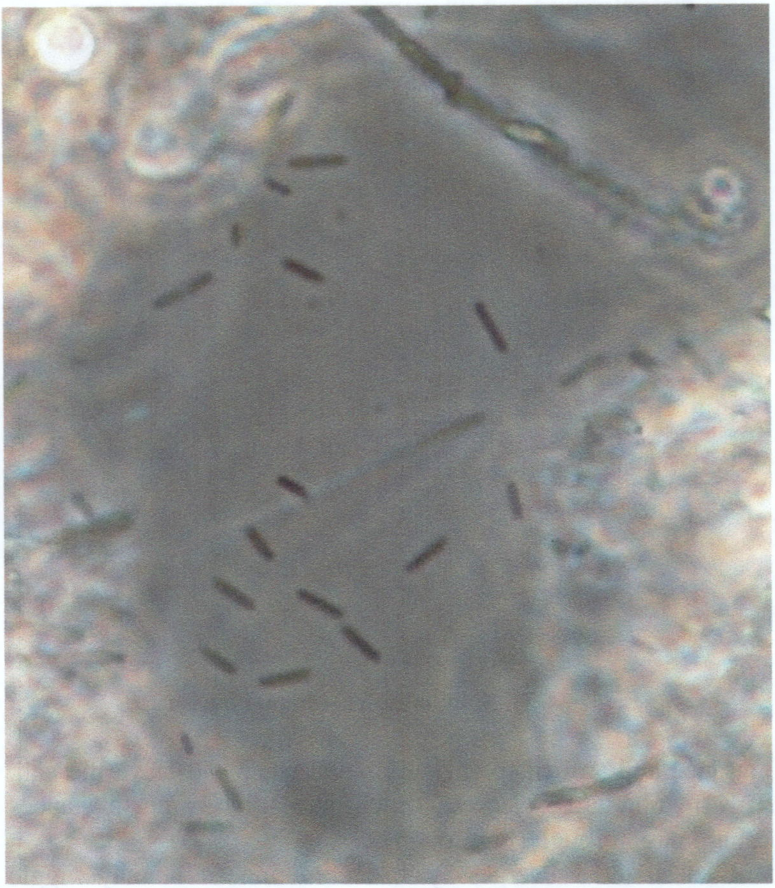

Figure 4-2 Bacilli, rod-shaped bacteria.

Spiral bacteria are never straight. Some, called *vibrios*, look like commas; *spirilla* (Figure 4-3) have rigid bodies and are shaped like corkscrews. *Spirochetes* are also corkscrew shaped but with more flexible bodies (Figure 4-4).

Many bacteria develop an outer sticky substance called a *slime layer*. This material is made inside the cell and is excreted to the outside. This slime layer enables the bacterial cells to stick to one another. A typical bacterial cell is also surrounded by a semirigid cell wall that gives it its shape and protects its internal parts. Just inside the cell wall is the cell membrane. This functions as an

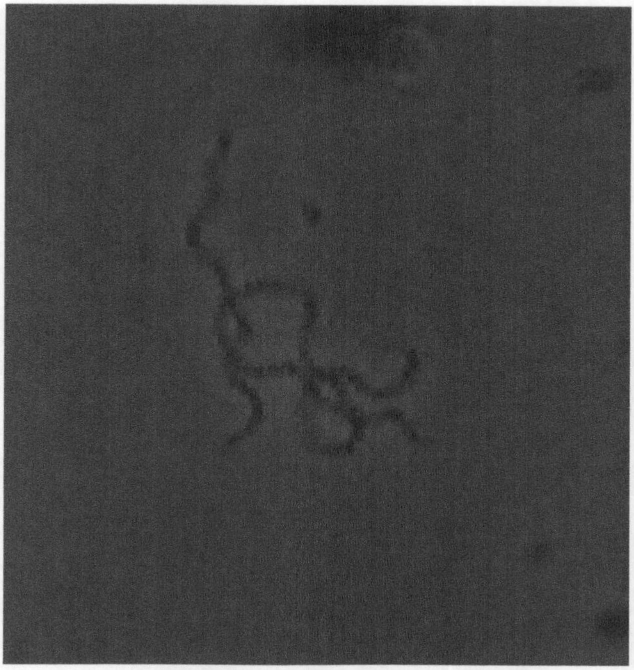

Figure 4-3 Spirilla, spiral-shaped bacteria.

entrance and an exit for materials that enter and leave the cell. This membrane is selective, which means that some substances cannot pass through (Figure 4-5).

Bacteria have several ways of locomotion. Some may wiggle or squirm; some glide through their medium; others propel themselves through the use of *flagella*. Flagella are long whip-like structures that enable bacteria to move through the water. Flagellated bacteria can move in one direction, they can tumble, and they can move in reverse.

Most bacteria are structured to function very well in the activated sludge environment. They are nature's ideal employees. They can use most of the organic material common to sewage. They are designed to swim to where the organic nutrients are, multiply to meet the demand, and they know how to settle out only after the job is done.

Bacteria can use only soluble nutrients, so dissolved organic material with smaller molecules pass easily through the cell

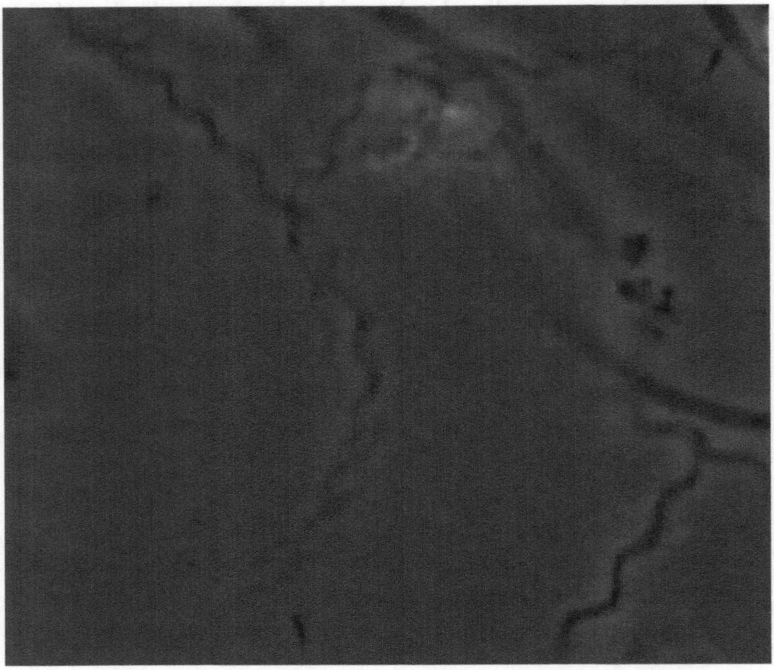

Figure 4-4 Spirochete, spiral-shaped bacteria.

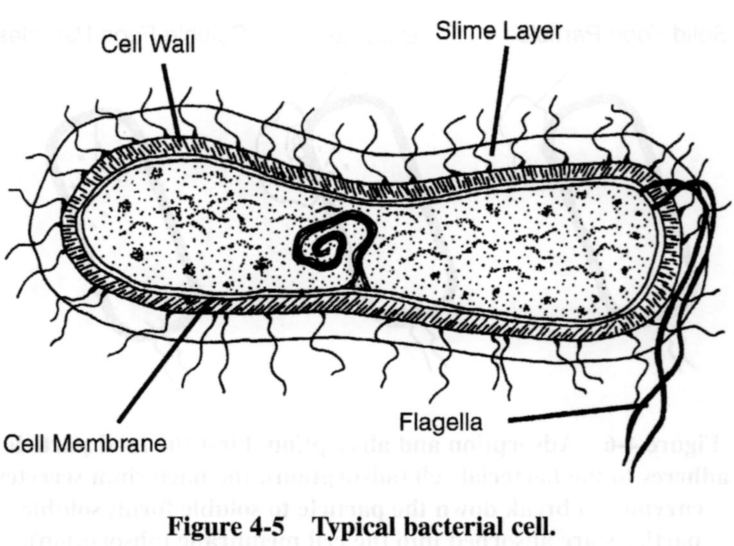

Figure 4-5 Typical bacterial cell.

membrane. Larger, undissolved particles have to be broken down to much smaller units before they can be transported into the cell. In this case, larger particles adhere to the sticky slime layer of the cell wall. This process is called *adsorption*. Next, the cell membrane secretes enzymes that cause chemical reactions that break the large particles down into smaller units that can be transported across the cell membrane and into the cell. This is the process of *absorption* (Figure 4-6).

Most bacteria in activated sludge use organic compounds. Some, however, use inorganic compounds such as iron, sulfur, and nitrifying bacteria. Sewage from homes, industries, and storm water that enters the wastewater treatment system contains many types of organic substances. Bacteria require essential elements such as carbon, nitrogen, phosphorus, sulfur, and water to build cell components. Other nutrients such as carbohydrates, sugars, and fats are used mostly for energy. Bacteria also need substances like magnesium, calcium, potassium, iron, and chloride. Although these substances are generally found in most municipal wastewaters, bacteria cannot use these substances without enzymes.

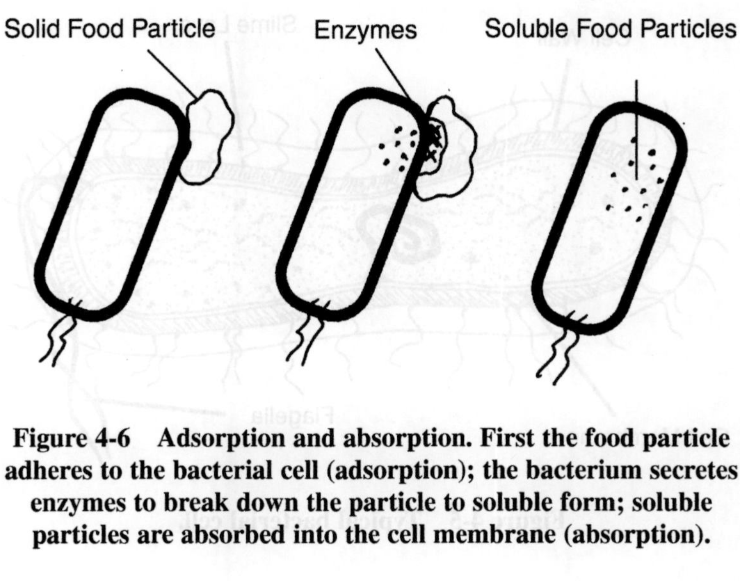

Figure 4-6 Adsorption and absorption. First the food particle adheres to the bacterial cell (adsorption); the bacterium secretes enzymes to break down the particle to soluble form; soluble particles are absorbed into the cell membrane (absorption).

Enzymes

Like all living organisms, bacteria carry out a process called *metabolism,* the sum of all the biochemical reactions that occur in a living cell. These reactions are necessary for life, growth, reproduction, and even death. Enzymes are proteins that trigger biochemical reactions in living organisms. All living organisms can produce thousands of enzymes. Some, called *exoenzymes*, perform their tasks outside the cell wall; others, called *endoenzymes,* perform their tasks inside the cell.

Enzymes are proteins and life process chemicals that can be affected by heat, acid, and alcohol and destroyed by toxic substances. Enzymes are very efficient. Under favorable conditions they can increase the rate of biochemical reactions 10 billion times more rapidly than reactions without enzymes. Enzymes are *catalysts*, which means they help biochemical reactions to occur more rapidly without being consumed in the process. Thousands of enzymes in one cell must function properly or the cell will stop functioning. The function of enzymes may be altered by changes in the environment in the wastewater treatment system. Environmental factors such as the concentrations of enzymes, nutrients, temperature, pH, and the presence of chemical substances that serve as enzyme inhibitors all affect enzyme function.

Enzymes also help determine the specific traits of each organism. They distinguish the aerobic bacteria from the anaerobic bacteria; they distinguish those that prefer acidic environments from those that prefer basic environments; and those that prefer heat from those that prefer midrange temperatures. Without enzymes, microorganisms cannot break down nutrients or perform cellular processes.

Commercial Enzymes

Enzymes are commercially available for use in the wastewater treatment system. Commercial enzymes can be useful when the wastewater is rich in a certain substance from a particular source. For example, the enzyme proteinase can be used to help break down protein-rich waste from meat packing industries. In this situation the enzyme should be added before the aeration basin so it

can break the protein-rich compounds into smaller units that the bacteria can easily consume. The higher the concentration of enzymes, the faster the breakdown of the substrate.

Super Bugs

Super bugs are selectively or genetically engineered to do one thing better than the normal bacteria. They can produce extra enzymes that break down certain substrates. Genetically altered bacteria can significantly expedite the breakdown of certain substances. For example, genetically altered bacteria are available for use in oil spills.

Most municipal systems do not need enzyme supplements or super bugs under normal operating conditions. No super bug alone can break down all the substances found in wastewater. They can be useful, however, when dealing with extremely difficult-to-treat waste products. The bacteria normally found in the treatment system can treat wastewater from natural sources.

Growth in Activated Sludge

As long as enough nutrients are available, bacteria can multiply very rapidly by splitting into two identical cells. Some can divide in two in only 11 minutes. Many can double in 20 to 30 minutes. In fact, if enough nutrients were available, *Escherichia coli* bacteria could produce a mass of bacteria greater than the total mass of Earth in 3 days. Bacteria are much like us, if lots of nutrients are available, they use them mostly for growth and multiplication. Although we do not split in two, we do increase in bulk. Bacteria, on the other hand, will simply continue to increase in number.

Let's consider the flow of mixed liquor through the aeration basin. Wastewater is mixed with the microorganisms in the return sludge at the beginning of the aeration basin, where the level of food is quite high. Because an abundance of food is available, bacteria are growing, motile, and extremely active. They have flagella, which enable them to swim to find available food. At this point the bacteria are multiplying rapidly. Essential to good

treatment is the formation of sludge floc. Actively multiplying bacteria cannot form sludge floc.

However, as the level of food decreases, the bacteria begin to conserve energy and they lose the flagella. Since little food is available, they do not waste energy trying to swim around to find it. Next they begin to form a thick outer slime layer. As they bump into each other, they stick together. At first they form small clumps then masses large enough to settle (Figure 4-7).

The bacteria must remain in the aeration basin long enough to remove most of the food so floc can form. The amount of time that the bacteria reside in the basin is called the *detention time*. Detention time can also be defined as the amount of time it takes for the water to flow from one end of the basin to the other. If the detention time is not long enough, the amount of food remaining will still be high, the bacteria will actively move and reproduce, and the discharge into the settling basin will be full of free-swimming, dispersed bacteria. The result will be poor settling sludge and a turbid effluent. If, however, the detention time in the aeration basin is long enough to allow the bacteria to consume most of the food, they will lose flagella, form a sticky slime layer, clump together, and form floc.

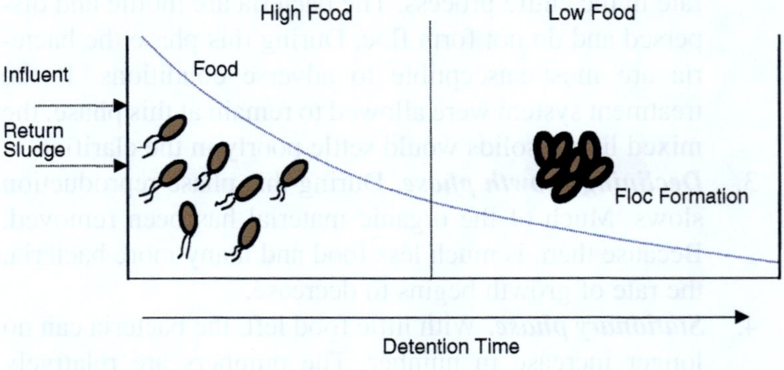

Figure 4-7 When the level of food is high, bacteria grow rapidly and disperse. As the level of food decreases, bacteria lose their flagella and begin to form a thick slime layer that enables them to clump together to form floc.

Growth Curve

The growth cycle for bacteria involves five phases:

1. ***Lag phase.*** During this phase the bacteria become acclimated to the environment. They take time to "taste the wares" and begin to produce the necessary enzymes to break down the substances detected in the wastewater. The lag phase can last from 1 hour to several days. There is no growth or reproduction during this phase, because bacteria need a time for adjustment. It is difficult for the bacteria to become acclimated in a treatment system that has variable loads. They need time to develop the proper enzymes. If the loading changes drastically from moment to moment, the bacteria cannot adjust. Actually, the lag time is extended. This means that there is a longer period of time before the bacteria can begin to consume the organic material from the water. Ideally, in such cases, an equalization tank would stabilize the loading.
2. ***Accelerated growth phase.*** Once the bacteria have produced enough of the necessary enzymes to help break down the organic load, they begin to grow rapidly. Here the bacteria constantly divide and the population doubles. The cells are active and the food is reduced at the fastest rate in the entire process. The bacteria are motile and dispersed and do not form floc. During this phase the bacteria are most susceptible to adverse conditions. If the treatment system were allowed to remain at this phase, the mixed liquor solids would settle poorly in the clarifier.
3. ***Declining growth phase.*** During this phase reproduction slows. Much of the organic material has been removed. Because there is much less food and many more bacteria, the rate of growth begins to decrease.
4. ***Stationary phase.*** With little food left, the bacteria can no longer increase in number. The numbers are relatively constant. The number of dying cells is about equal to the number of reproducing cells. In this phase the cells begin to agglomerate and conserve energy. The bacteria lose their flagella and begin to form a thicker slime layer.

5. ***Death phase.*** In the death phase, the number of deaths exceeds the number of new cells. The nutrients are depleted and toxic waste products, which create a very unfavorable environment, increase. If the detention time in the aeration basin is too long and allows most bacteria cells to enter the death phase, the once healthy floc solids will begin to break apart. The solids will not settle well in the clarifier, and this will produce an unacceptably turbid effluent.

Bacteria in Wastewater Treatment

Among all the types of microorganisms involved in the wastewater treatment plant, bacteria are the most numerous and the most important. Most bacteria in the activated sludge process that form floc are gram-negative and rod-shaped. The operator's primary role is to maintain conditions in the treatment system that favor the growth of the bacteria that are best suited to remove the organic material from the wastewater. An operator can control some conditions that affect their growth, but not all.

Temperature directly affects the rate at which biochemical reactions occur. Biochemical reactions occur much faster in warmer temperatures. Therefore, fewer microorganisms are required to remove nutrients when the temperatures are warmer. Conversely, in colder temperatures, more microorganisms are required to do the same amount of work than is required in warmer temperatures. Most microorganisms associated with sewage perform best at temperatures in the 20°C range.

Bacteria require essential elements and nutrients to build cell components and remove other substances from the wastewater. Most essential nutrients are commonly found in municipal wastewater. However, wastewater from most industries lacks many essential nutrients for proper cell growth and multiplication. When critical nutrients are absent from the wastewater, the bacteria will not perform well. We will discuss the effects of nutrient deficiency later in this chapter. Operators have very little control over the discharges from homes and even storm water runoff that enters the treatment system. The operator can, however, add

nutrient supplements when essential nutrients like phosphorus or nitrogen are lacking.

Bacterial enzymes (proteins) depend on pH. The optimum pH for wastewater microorganisms is in the neutral range of 6.5 to 8.5. Fluctuations or rapid changes in pH will alter or denature the structure of the enzymes. Filamentous bacteria, fungi, and yeast can compete better than floc-forming bacteria in low pH conditions.

Since aerobic bacteria are the primary agents in activated sludge, they require oxygen. The operator has some control over the amount of oxygen available to the microorganisms. The rate of oxygen use can be correlated with the amount of available food. When the level of food is high, bacteria actively grow and use oxygen rapidly. As the level of food declines, there is less activity and the requirement for oxygen decreases. Bacteria require 0.1 to 0.3 mg/L of dissolved oxygen (DO) to function properly. However, a DO level of 2 mg/L should be maintained throughout the aeration basin. Activated sludge floc is made up of bacteria, so for bacteria in the center of the floc to receive enough oxygen, a DO of 2 mg/L should be maintained in the bulk fluid that surrounds the floc. If too little DO reaches the center of the floc, the bacteria in the center will die. The floc solids will begin to break up and may begin to develop filamentous bacteria that thrive under low DO conditions.

Floc Formation

The structure and size of the floc are important in maintaining the settleablility of the sludge. Early in the treatment process very little new sludge is formed because bacteria are motile and dispersed. As the sludge begins to age, small clumps and chains of bacteria begin to stick together to form floc until eventually they are large enough to settle. As the floc ages, many nitrifying bacteria establish on the surface because many chemical ions, including ammonia, accumulate on the surface of solid objects.

Proper floc formation is critical to the wastewater treatment process. The following factors can adversely affect floc formation:

- Sludge age
- Slug discharges

- Toxicity
- Surfactants
- Excessive shearing

If the sludge is too young, the bacteria will remain active and will not clump together to form floc. They must be allowed to remain in the aeration basin long enough to experience low-food conditions so they can develop the slime layer needed for proper floc formation. Slug discharges have a similar effect. In the presence of excess food, the bacteria will not form floc.

Some toxic substances may kill the microorganisms; others may retard their ability to remove nutrients from the wastewater. For example, the toxic impact from petrochemical poisoning will cause bacterial cells to build a protective wall. The wall prevents the cells from sticking together to form floc. Thus, they do not settle well in the clarifier.

Once the bacteria in the floc are dead, the floc will break apart. Dead bacteria do not form into floc. Toxic substances generally have a greater effect on higher life forms, like stalked ciliates and rotifers. They may have the advantage in competing for food because they have the ability to draw in large amounts of the wastewater. On the other hand, for the same reason, they are at a disadvantage in the presence of toxic substances.

Surfactants, which are substances that reduce the tension on the surface of the water, make water slippery. Although surfactants will not prevent the bacteria from developing a slime layer, they can interfere with the stickiness of the slime and prevent the bacteria from sticking together. Additionally, excessive shearing caused by coarse air diffusers will cause floc to break apart.

Other Parameters That Affect Bacteria

Food to Microorganism (F/M) Ratio

The F/M ratio compares the amount of available food with the number of microorganisms in the aeration basin. A high F/M ratio means there is a great deal of food and only a few microorganisms. When the F/M ratio is high the bacteria are active and dispersed

and multiply rapidly. They do not form floc. Operating the wastewater treatment system with a high F/M ratio will probably yield a turbid effluent and a poor settling sludge. A low F/M ratio means the opposite. This means there are many microorganisms with only a limited amount of food. Only when food is limited do bacteria begin to develop a thicker slime layer, lose their motility, and begin to clump together to form floc.

How is the amount of available food determined? The F/M ratio should be determined using the amount of food that enters the aeration basin in relation to the amount of microorganisms in the basin. The amount of food is measured using the BOD or chemical oxygen demand (COD) test. The BOD test is an indirect measure of the amount of food. To more accurately measure the amount of food available to the microorganisms, the BOD should be measured using the wastewater that enters the aeration basin. If primary treatment is available, the primary effluent should be used.

As microorganisms consume the organic material in the wastewater, they use oxygen. The BOD test measures the amount of oxygen the microorganisms require as they consume the available food. If a great deal of food is available, the demand for oxygen will be high. Conversely, if little food is available, the demand for oxygen will be low. One drawback to this test is that the sample has to be incubated for 5 days. The concentration of oxygen in the water is measured before and after the 5-day period. The difference is the amount of oxygen the microorganisms required as they consumed the food during that period of time. The COD test measures the amount of oxygen that is used as chemicals in the wastewater break down to more stable forms. In this test, oxygen is released into the sample by potassium dichromate. As the chemicals break down to more stable forms, oxygen is used up. The oxygen demand is determined by the amount of potassium dichromate that is used up. Unfortunately, this test does not measure the amount of oxygen used by the microorganisms. So, why is it used? The amount of BOD can be estimated using the COD test. On average, the BOD is 50 to 70 percent of the COD. The advantage to this test is that it takes only 3 hours compared to the 5 days required for the BOD test.

How is the concentration of microorganisms determined? Generally, the mixed liquor volatile suspended solids (MLVSS) concentration is used to determine the concentration of microorganisms in the aeration basin. Remember, the mixed liquor in the aeration basin is a mixture of microorganisms and wastewater. A sample of mixed liquor is poured and captured onto a filter that is dried to determine the dry weight. This dry weight is the mixed liquor suspended solids (MLSS) concentration. The filter is then heated to about 550°C to burn off all the organic solids. The percentage that is burned off is called the volatile solids concentration or mixed liquor volatile suspended solids (MLVSS). This volatile portion is assumed to be mostly microorganisms. So, the amount of microorganisms is determined by measuring the MLVSS concentration.

How is the F/M ratio determined? The F/M ratio is the amount of food (as determined by the BOD or COD test) divided by the amount of microorganisms (as determined by the MLVSS test):

$$\frac{\text{lb BOD}}{\text{lb MLVSS}} = \text{F/M ratio}$$

The F/M ratio affects the performance of the microorganisms in the activated sludge. Understanding how the F/M ratio affects the microorganisms will give the operator a valuable process control tool. There is not an ideal F/M ratio that will work for all activated sludge treatment systems. Generally, a lower F/M (0.05 to 0.5) is acceptable.

Sludge Age

Sludge age refers to the average age of the sludge or microorganisms in the activated sludge system (aeration basin and clarifiers). For the purpose of this book it refers to the average time the microorganisms remain in the aeration basin, where bacteria come in contact with the incoming wastewater as a food source. There is an initial adjustment period, during which the bacteria assess the environment and develop the enzymes that will be required to break down the nutrients they have detected in the wastewater. Once this occurs, they begin to consume the nutrients and multiply

at a logarithmic rate. The bacteria are motile and active and do not clump together to form floc. Only by remaining in the basin long enough to deplete most of the nutrients do the bacteria begin to form the slime layer necessary to form floc. Enough bacteria must be allowed to remain in the aeration basin long enough to be exposed to conditions where food is limited. If the sludge is too young, this will not occur. On the other hand, if the sludge is too old, floc will begin to break apart and yield a turbid effluent.

Influent Wastewater Nutrient Composition

Like most organisms, bacteria require nutrients to grow and maintain cell functions. They need carbon, nitrogen, phosphorus, and sulfur in addition to various metal ions such as magnesium, calcium, iron, and copper. Most domestic wastewaters contain enough carbohydrates, proteins, and fats, which supply most of the nutrients that bacteria require. The ideal formula for determining the nutritional needs of microorganisms, expressed in a ratio of carbon:nitrogen:phosphorus (C:N:P), is 100:10:1. If BOD is the carbon source, the ratio will be more like 500:10:1. If any of these nutrients are lacking, operational problems such as slime bulking and foaming may result.

Slime Bulking and Foaming

Slime bulking occurs when bacteria produce an excess of a substance called *exocellular lipopolysaccharide*, better known as *slime*. This occurs primarily in floc-forming, gram-negative bacteria and is a function of the cell wall structure.

The cell wall on a gram-positive bacterium is made up of three primary components: a polysaccharide layer, the cell wall, and the cell membrane. But gram-negative bacterium has five primary components: a lipopolysaccharide layer, a phospholipid layer, a lipoprotein layer, the cell wall, and the cell membrane. The lipopolysaccharide layer on a gram-negative bacterium is made up of a gelatinous, polymer-like substance that resists the purple dye of the Gram stain, hence the reason it is gram-negative (Figure 4-8).

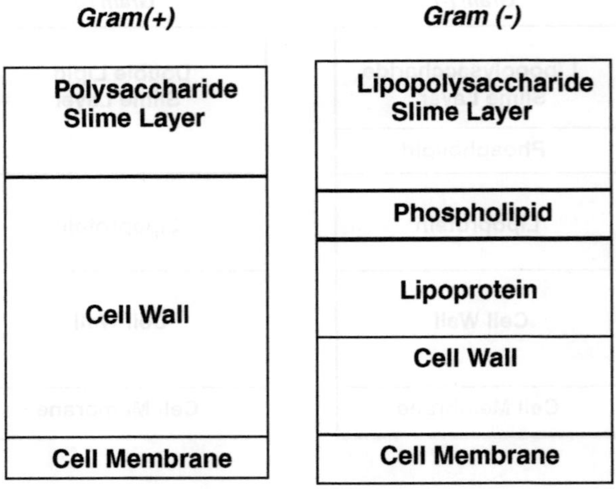

Figure 4-8 Cross-section of bacterial cell wall that shows the difference between the gram-positive and gram-negative cell wall structures. Usually slime bulking is caused by the gram-negative bacteria.

Three main elements are required to develop this gram-negative cell wall: nitrogen, phosphorus, and sulfur. If any of these elements are lacking, the cell wall does not develop properly. For example, phosphorus is required to develop the phospholipid layer, and nitrogen is required to develop the lipoprotein layer.

If there is a deficiency of phosphorus, the phospholipid layer will not develop properly. Instead of a phospholipid layer, there will be a layer that is predominantly made up of lipid or fat. The cell wall now has its lipopolysaccharide layer (which is already fat and slimy), along with an additional layer of fat (Figure 4-9). If there is also a deficiency of nitrogen, the cell will have a significant addition to the layers of fat or triple-lipo (Figure 4-10). This extra-thick layer of fat and slime interferes with the development

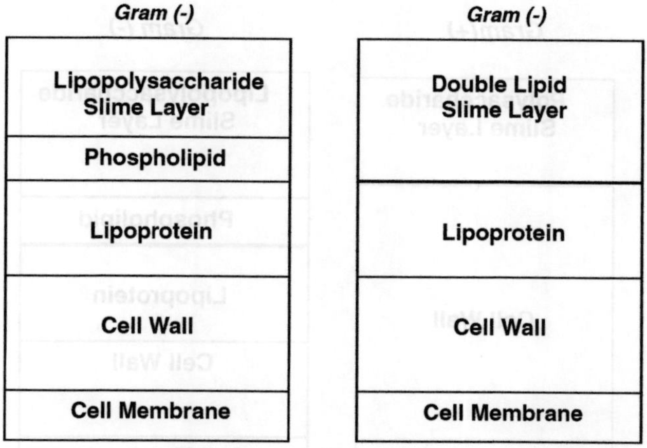

Figure 4-9 When phosphorus is limited, the phospholipid layer does not form properly. Instead an additional layer of fat is added to the slime layer.

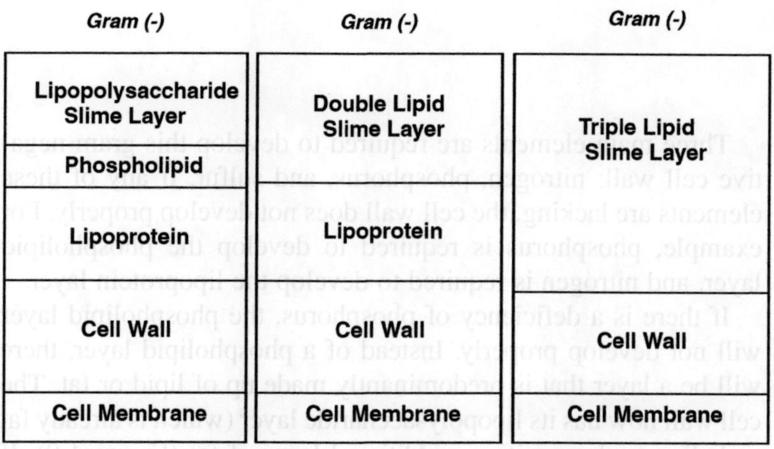

Figure 4-10 When nitrogen is deficient a third level of fat is added to the slime layer.

of floc. The bacterial cells do not compact closely together by a thin layer of slime; they are now spaced farther apart by a much thicker layer of fat and slime. This extracellular lipopolysaccharide resists water and causes the sludge to remain in whichever position it is in and it will not settle. The extra slime layer may also interfere with the cell's ability to remove nutrients from the water because it becomes difficult for them to penetrate through the slime layer to enter the cell wall.

Chapter 5

Protozoa

Bacteria are the primary agents in the removal of organic nutrients from the wastewater treatment system, but they are boring to watch under the microscope (especially with the microscopes most operators can afford). Protozoa, on the other hand, are larger (5 μm to 1 mm) and come in a variety of shapes with a wide range of behaviors. They are more interesting to watch.

Bacteria make up about 95 percent of the microorganisms in activated sludge; about 3 percent are protozoa. Protozoa do, however, play a critical role in the treatment process by removing and digesting free-swimming dispersed bacteria and other suspended particles. This improves the clarity of the wastewater effluent. Like bacteria, some protozoa need oxygen, some require very little oxygen, and a few can survive without oxygen.

Why bother looking at protozoa? The types of protozoa present give us some indication of treatment system performance. While in a steady state system, there is a natural progression in the dominance of the different protozoan species; activated sludge systems are far from steady state. The type and concentration of organic nutrients constantly change, as do factors such as temperature, amount of available oxygen, and food:microorganism ratio. Operators should not rely solely on the microscopic identification of protozoa. However, the relative dominance of the different protozoa species does give us an understanding of treatment system conditions. Many times a sudden change in the number and type of protozoa in the system can predict problems if adjustments aren't

made. How protozoa move, eat, and compete for food will determine which species will dominate at any given time.

Types of Protozoa

Most protozoa are aerobic, and all live in areas with a large water supply. There is an enormous diversity in types of protozoa, and they are extremely varied in appearance and lifestyle. Some of the smaller protozoa, like bacteria, take in soluble nutrients by absorption through the cell membrane. Others have specialized structures or mouth-like openings and feed on other microorganisms such as bacteria and algae or other solid matter. Protozoa are also single-celled microorganisms and are usually classified based on how they move. For the purpose of studying activated sludge, we will classify them as follows:

- Amoebae
- Flagellates
- Ciliates
- Free-swimming ciliates
- Crawling (grazing) ciliates
- Stalked (sessile) ciliates

The wastewater operator need not know how to precisely identify all the species of protozoa. It is helpful, however, to classify them into one of the five categories listed above because each species of protozoa dominates under different conditions.

Amoebae

Amoebae are the most primitive type of protozoa. They feed on small organic particulates but will also eat algae, bacteria, and other small protozoa. The amoeba is not just a shapeless blob waiting for food to draw near. It can sense a nearby concentration of food and will be drawn to it. Although the amoeba can absorb soluble nutrients, it is more drawn to particulate matter.

Amoebae contribute very little to the overall treatment of wastewater. They move slowly and cannot compete with bacteria

or other protozoa for food. Amoebae can only dominate early in the treatment process, when the amount of food is quite high. They begin to die as the amount of food decreases and the competition for food increases. Thus, when a sample is collected at the end of the treatment period, usually at the discharge end of the aeration basin, large numbers of amoebae are normally not present in the sample.

Identifying Amoebae Under the Microscope

Amoebae are relatively clear and move slowly. Sometimes they do not move at all. They can sit dormant for hours. They do very little to bring attention to themselves, which makes observing them under the microscope difficult. Amoebae should be observed under high power with a phase contrast condenser with as little light as possible.

Amoebae can be categorized into two types: the naked amoeba and the testate amoeba (which has a shell). The naked amoeba is most commonly seen in science books and depicted in horror movies. It moves slowly by extending lobe-like projections called *pseudopods*. Whenever it senses nearby food, it begins to extend its pseudopods until they have completely enveloped the food and form a cavity or *vacuole*. Enzymes are then secreted into the vacuole to break down the food into soluble form so it can be absorbed into the cell (Figure 5-1). The vacuole will move slowly down the body of the cell and decrease in size as the nutrients are broken down and are absorbed into the cell.

The testate amoeba has a shell or test. It produces its test by either secreting it or by building it from particles collected as it travels. The way the test is constructed is a useful way of identifying the testate amoeba, along with whether the pseudopodia are filose (threadlike), lobose (thick and irregular), or reticulose (straight). Testate amoeba are sometimes difficult to identify. Many are mistaken for a piece of organic material. Careful and patient observation is required to properly identify them.

The operator need not identify each type of amoeba. Determining the number of amoebae in the sample compared to the other types of protozoa is helpful. *Proteus* is a naked amoeba (Figure 5-2). It is the species most likely to be seen when

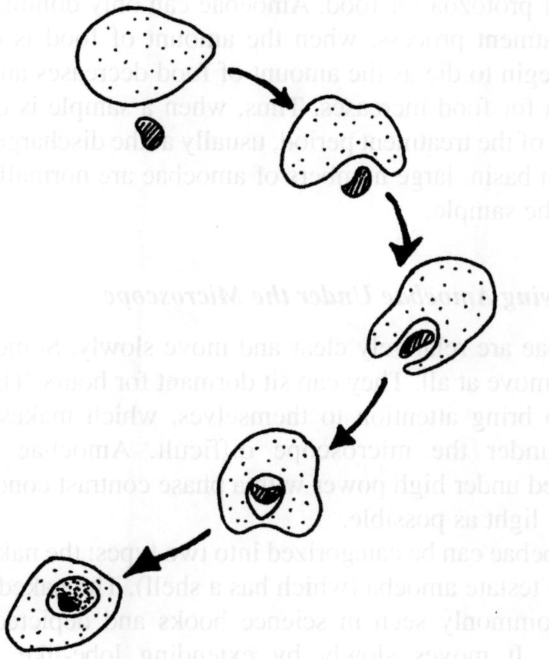

Figure 5-1 Amoebae feeding sequences.

looking for amoebae. This amoeba is completely exposed and vulnerable to environmental conditions. Its internal structures and pseudopodia are clearly seen. The testate amoebae are often overlooked or mistaken for a piece of debris. *Arcella* (Figure 5-3) and *Arcella discoides* (Figure 5-4) are both testate amoebae. The shell provides protection and enables them to survive in environments that are too harsh for the naked amoebae.

Flagellates

A flagellate is so named because it possesses a whip-like structure that helps pull the organism through the water. A flagellate is usually quite small and has one or two flagella. It has a tough outer membrane and can ingest food through a specialized mouth called a *cytostome*. The flagellum found in the early life stage of a bacterium is much different from that of the flagellate. A

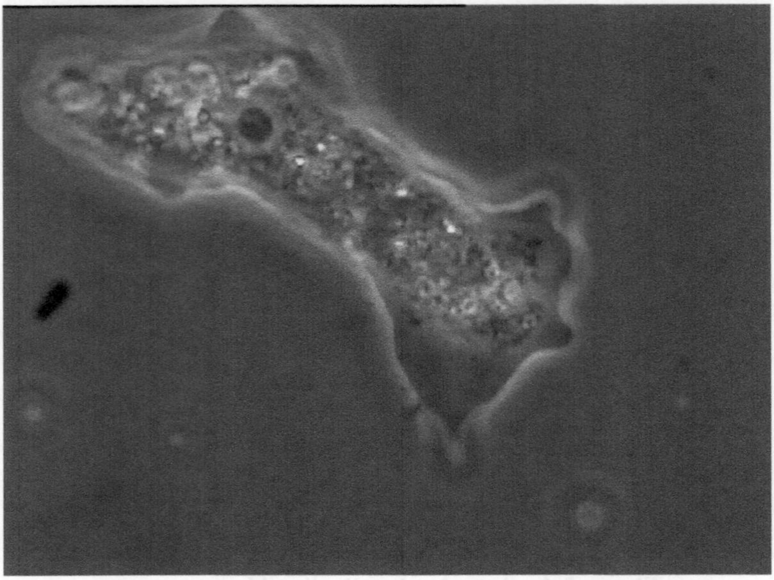

Figure 5-2 Amoeba *proteus* (1000× phase contrast).

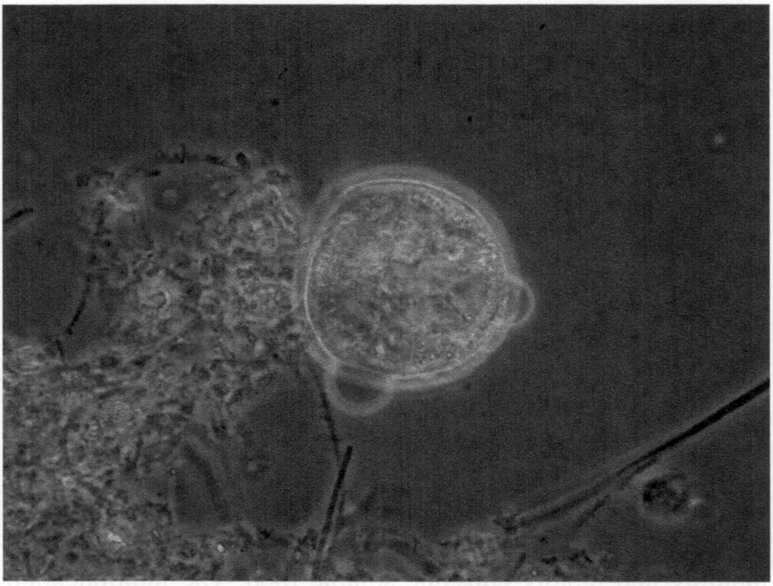

Figure 5-3 Amoeba *Arcella* (1000× phase contrast).

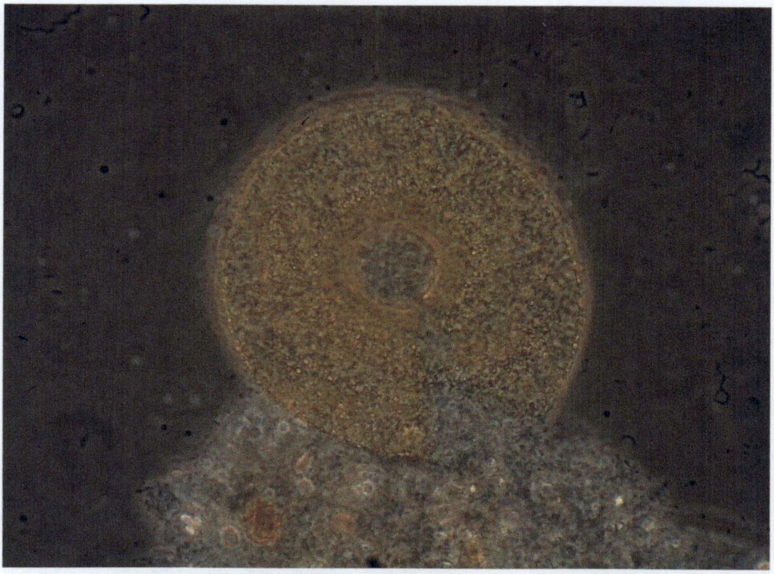

Figure 5-4 Amoeba *Arcella discoides,* a testate amoeba (1000× phase contrast).

bacterial flagellum is like a small device that moves by rotating the shaft, the part where the flagellum is attached to the cell. In a bacterium, it works like a little propeller. In flagellates, the flagella are composed of small tubules of protein called *microtubules*.

Amoebae prefer to feed on small particulates; flagellates thrive in waters that are rich in organic nutrients, especially where there is much decay. Some groups of flagellates feed on bacteria and small algae. However, the groups that thrive in the activated sludge treatment system seem to feed primarily on soluble organic nutrients, so they can only dominate early in the treatment process when the amount of nutrients is quite high. Flagellates compete with bacteria for soluble nutrients; therefore, they can only dominate when the bacteria population is still relatively small. While the bacteria acclimate to the environment, flagellates are free from any real competition for food. However, after the bacteria's initial lag phase comes an accelerated growth phase. Bacteria begin to multiply and rapidly consume soluble food. The multiplication

rate for bacteria is 15 times faster than that of the flagellates, so they are not without competition for long.

Identifying Flagellates Under the Microscope

Flagellates can be seen in a variety of shapes and sizes. They are the most widespread of the protozoa. Most are relatively small and easily distinguished compared to the other protozoa commonly found in activated sludge. Bodo is the most common flagellate found in activated sludge (Figure 5-5). Its short, capsule-shaped body is often seen spiraling through the fluid. The whip-like structure, called a flagellum, is used for locomotion and feeding. Some flagellates like *Spongomonas* are not very motile but live in colonies held together by a gelatinous matrix (Figure 5-6).

Ciliates

Ciliates are either completely or partially covered with short, dense hairlike structures called cilia, which get their meaning from the

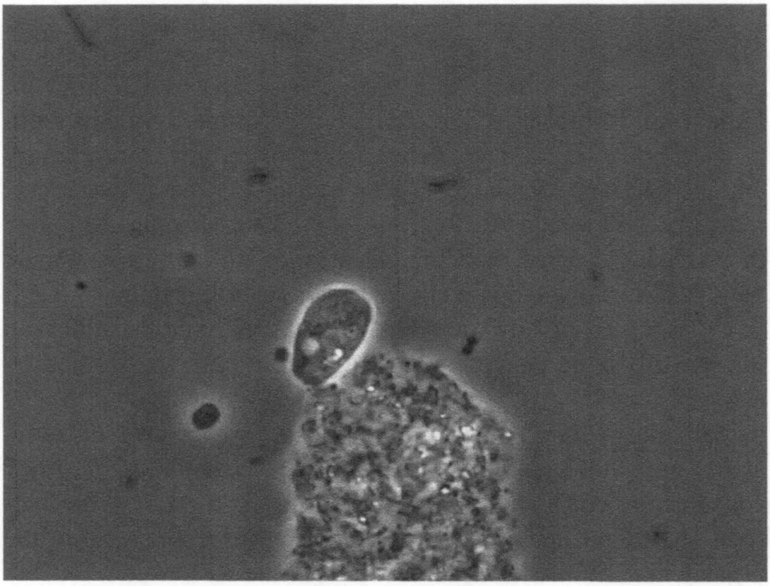

Figure 5-5 Flagellate Bodo (1000× phase contrast).

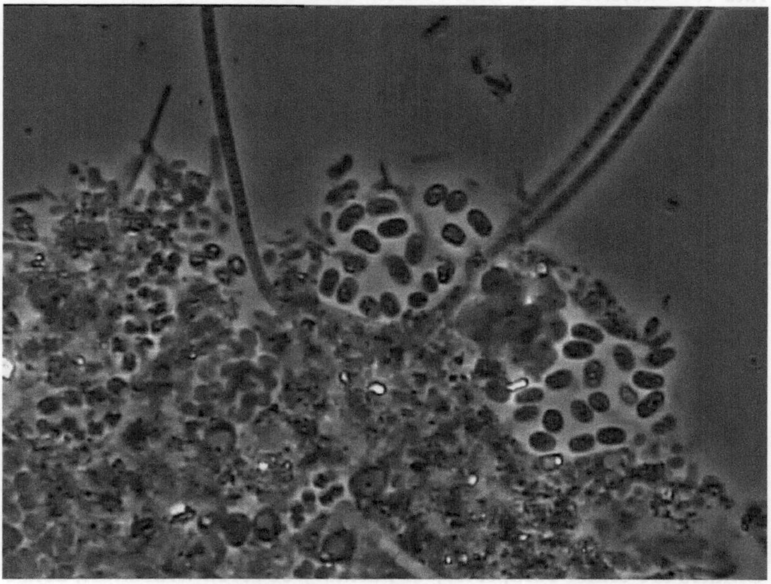

Figure 5-6 ***Spongomonas,*** **a colony of flagellates held together in a gelatinous matrix (1000× phase contrast).**

Latin word for eyelash. The cilia are used primarily for locomotion. Although ciliates actively capture their prey, the cilia are arranged in such a way that they can be used to generate water currents that will sweep the food into the ciliate's mouth. Although they are single-celled microorganisms, they still perform complex biological functions. Some ciliates reach 2 cm in length.

Ciliates feed mostly on bacteria, algae, and yeast, so they do not contribute to the overall treatment of the wastewater by actively removing organic material. They do contribute to the overall clarity of the water by removing suspended bacteria, algae, and other small microorganisms. Bacteria multiply as they consume organic material; ciliates multiply as they consume bacteria. An abundance of ciliates usually indicates that most of the organic material has been removed from the wastewater.

About 7,500 species of ciliates are generally classified based on the way the cilia are arranged. For the purpose of studying activated sludge, we will look at three categories of ciliates. They are categorized based on their feeding habits and their ability to compete for food.

Free-Swimming Ciliates

These are generally uniformly covered with cilia. They move by swimming and, in many instances, spiraling through the water. Some ciliates, like the *Paramecium*, can back up and change directions if they encounter a solid object. Since free dispersed bacteria are their principal food, free-swimming ciliates begin to dominate when the numbers of bacteria are quite high. Bacteria reach their peak after most of the soluble nutrients have been depleted. Flagellates cannot compete with bacteria for the organic nutrients, but free-swimming ciliates do not have to compete with the bacteria. Free dispersed bacteria give them the advantage. In the presence of an abundant food source, bacteria are motile and multiply actively. Free-swimming ciliates move quickly through this environment and consume as many bacteria as they can. Once the organic nutrients become limited, the bacteria will no longer swim freely, but will begin to lose the flagella, form a sticky slime layer, and clump together to form floc.

Identifying free-swimming ciliates under the microscope. Free-swimming ciliates use their cilia primarily for locomotion. They move quickly through the fluid and sweep bacteria into an oral groove. They are easily distinguished from the crawling ciliates because they generally live in the fluid. Something that swims quickly past your field of view in the microscope is probably a free-swimming ciliate. The *Litonotus* is known as a graceful swimmer (Figure 5-7). It is rarely observed within the floc. Paramecia are faster swimmers and are often seen darting in and out of floc particles as well as spiraling through the fluid (Figure 5-8). *Coleps* are known as the armored tanks of the protozoa world because their bodies are protected by many small plates and spines (Figure 5-9). They are also scavengers, feeding on the dead remains of other protozoa as well as bacteria.

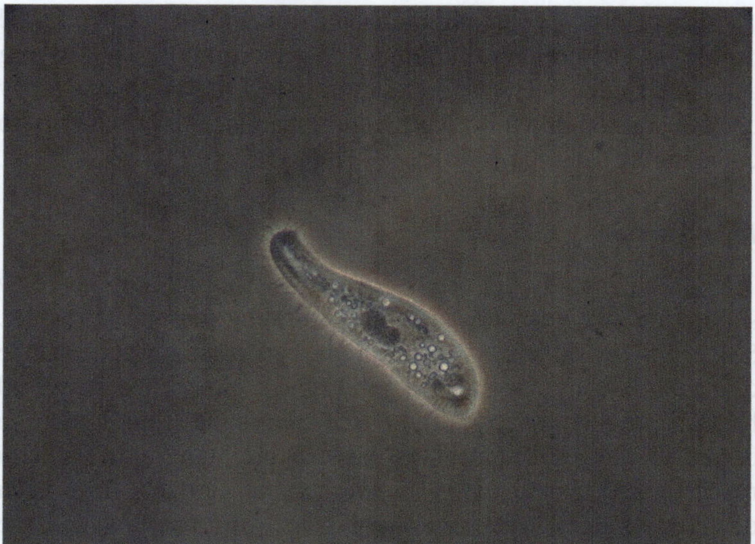

Figure 5-7 ***Litonotus,*** **a free-swimming ciliate (1000× phase contrast).**

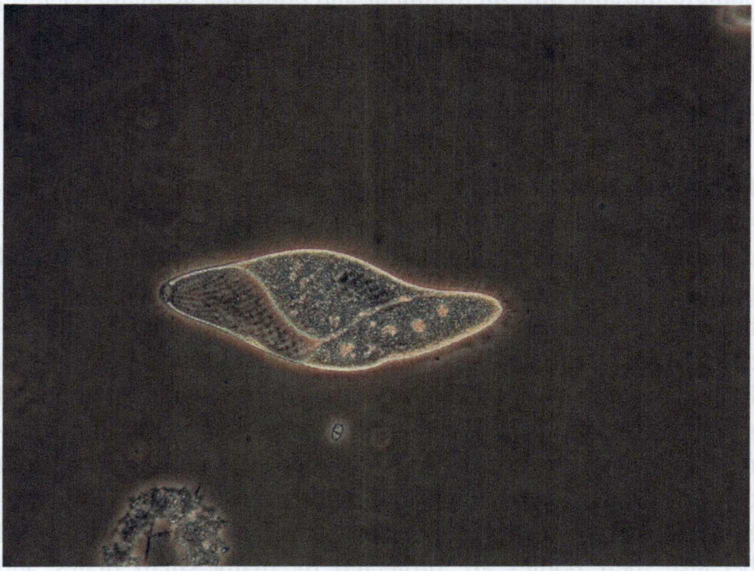

Figure 5-8 ***Paramecium,*** **a free-swimming ciliate (1000× phase contrast).**

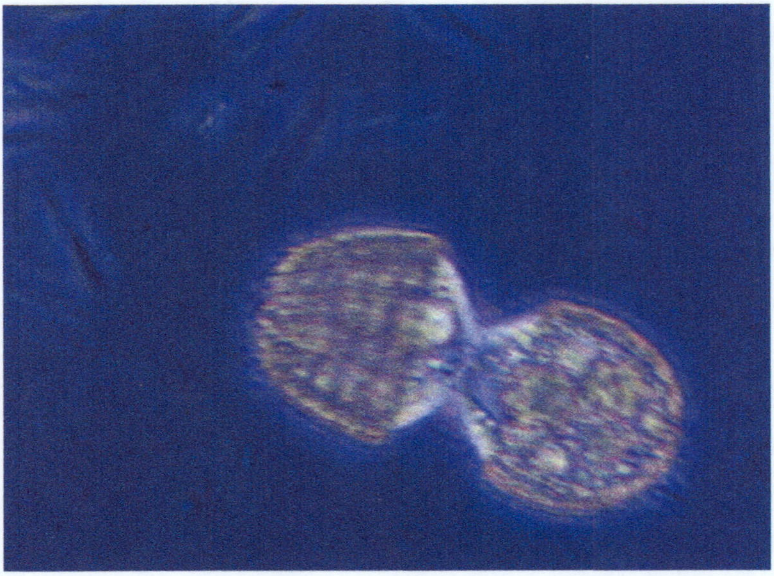

Figure 5-9 *Coleps* dividing, a free-swimming ciliate (blue filter 1000× phase contrast).

Crawling (Grazing) Ciliates

These are common in activated sludge. Their dominance in the system indicates good treatment conditions. Crawling ciliates begin to gain dominance after most soluble nutrients have been removed. During this time most of the dispersed bacteria have begun to clump together to form floc.

Identifying crawling ciliates under the microscope. Crawlers have cilia that are organized into structures called *cirri,* which are several cilia joined together to form a tuft or leg. Instead of swimming, the crawling ciliate has cirri on its underside that function as coordinated legs that allow them to walk on surfaces. They use these legs to browse or crawl along solid floc particles and pick at any straggling bacteria they can find. Crawlers are poor swimmers. If left in the fluid, they grope around, turn in circles, and try to touch something solid so they can climb aboard and begin to crawl (Figure 5-10).

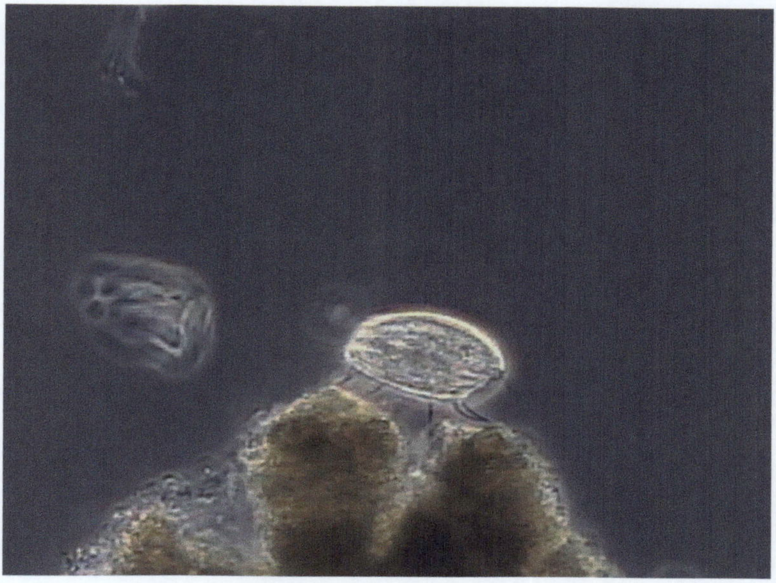

Figure 5-10 ***Aspidisca sulcata,*** **a crawling ciliate (1000× phase contrast).**

Stalked (Sessile) Ciliates

These have no cilia on their bodies other than those that fringe around the mouth ends. The cilia are used to create a current that brings food particles into the mouth. They feed mostly on suspended bacteria, algae, or smaller protozoa. They rarely swim freely, but can be found attached to almost anything. Some stalked ciliates grow as separate stalks; others branch into colonies that can be made up of 100 or more organisms. Sometimes stalked ciliates are enclosed in a vase-like sheath.

By the time stalked ciliates begin to dominate in the activated sludge, the bacteria have removed most of the organic nutrients; most of the free-dispersed bacteria have been removed by the free-swimming ciliates or have developed into floc. The stalked ciliate's advantage is that it can anchor itself while it uses its cilia to create a current that brings the food to its mouth.

Identifying stalked ciliates under the microscope. *Vorticella* (Figure 5-11) do not grow in colonies but are sometimes found in large groups with separate stalks. The stalks are tubular structures with central muscle fibers that cause the ciliate to contract whenever it is disturbed (Figure 5-12). Once the disturbance is past it will slowly extend itself again. *Vorticella* can detach themselves from their stalks whenever conditions become unfavorable, and swim freely to look for a more favorable environment. There they attach themselves and generate new stalks.

Carchesium (Figure 5-13), which looks a lot like *Vorticella*, is a colonial stalked ciliate. Several zooids, or heads, are attached to branching stalks. Each stalk has its own contractile muscle that contracts independently of the others and of the main trunk, which also has a contractile muscle. A *Carchesium* colony can contain hundreds of heads. The combined action of all the heads can consume a large quantity of suspended bacteria and algae.

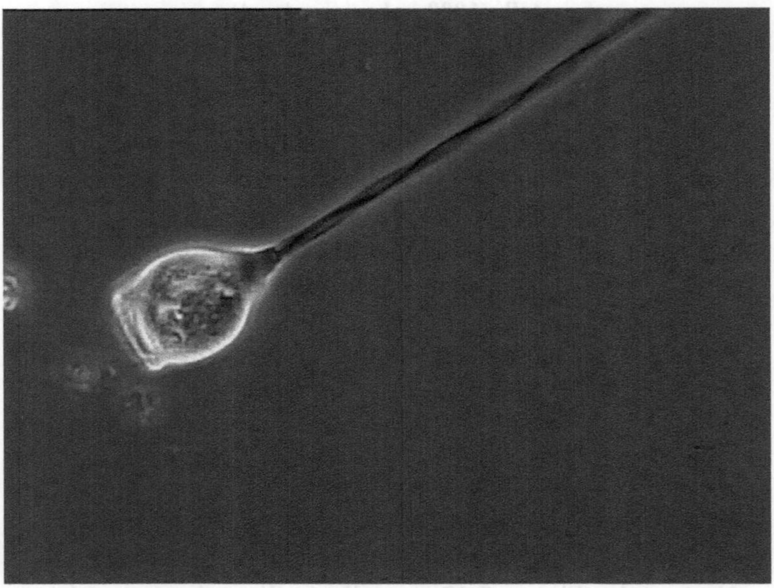

Figure 5-11 *Vorticella,* a stalked ciliate with a solitary head and contractile stalk. The stalk contains myoneme (muscle fiber).

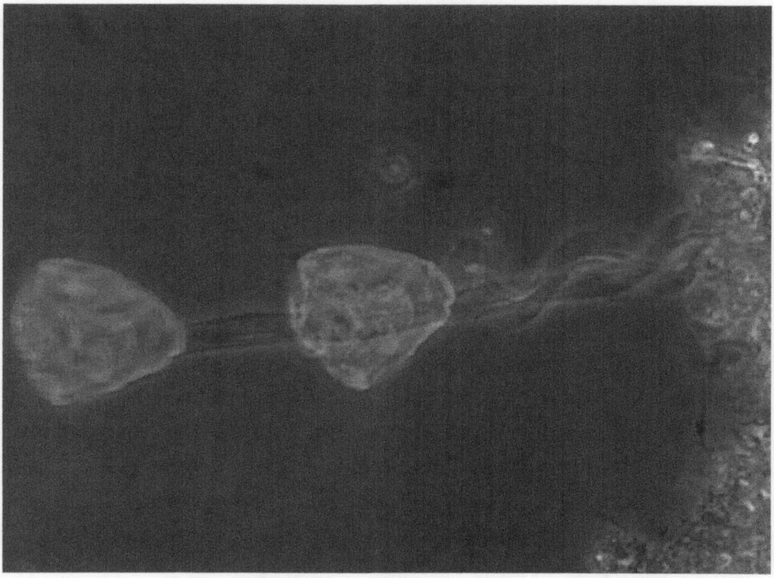

Figure 5-12 *Vorticella,* a contracting stalk (1000× phase contrast).

Epistylis sp. (Figure 5-14) is another type of colonial stalked ciliate. Similar to *Vorticella* and *Carchesium*, it has an inverted bell-shaped head but it is mounted on a branched noncontractile stalk.

Opercularia (Figure 5-15) also has a branched noncontractile stalk but grows in much smaller colonies that consist of 3 to 50 individual heads.

The *Stentor* (Figure 5-16) is one of the largest protozoa in water. It is a trumpet-shaped organism that is covered with short cilia. A crown of longer cilia, used to aid in feeding, surrounds the rim of the trumpet. *Stentors* can either swim freely or form colonies by secreting a sticky gelatinous substance that anchors them together. Any disturbance will cause the organism to contract into what looks like a blob.

Vaginacola (Figure 5-17) a stalked ciliate, is enclosed in a vase-like sheath. Often a pair lives together in one sheath.

Figure 5-13 ***Carchesium.*** **Each zooid, or head, can contract independently of the rest.**

Suctoria (Figure 5-18) has no cilia but, much like its name, it has lots of long tubular tentacles with sucker-like tips that suck the body fluids of its prey (Figure 5-19).

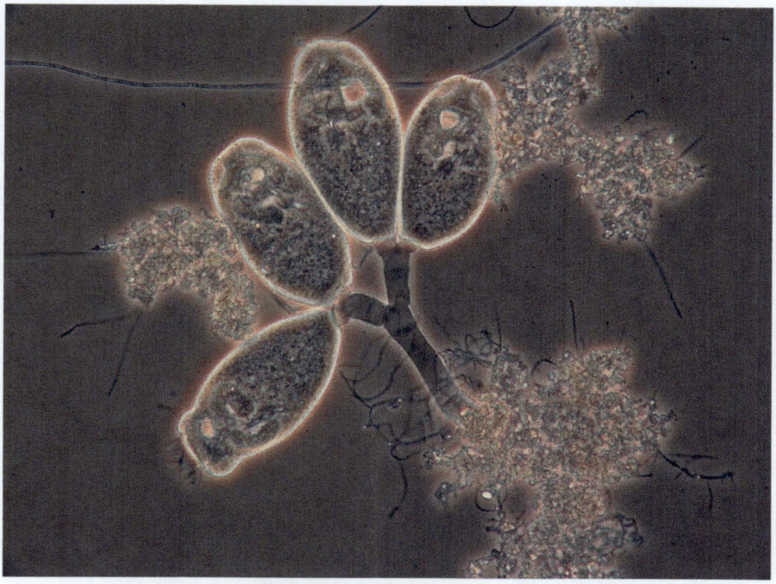

Figure 5-14 ***Epistylis* sp., a colonial stalked ciliate (1000× phase contrast).**

Figure 5-15 ***Opercularia;* no muscle fiber (400× phase contrast).**

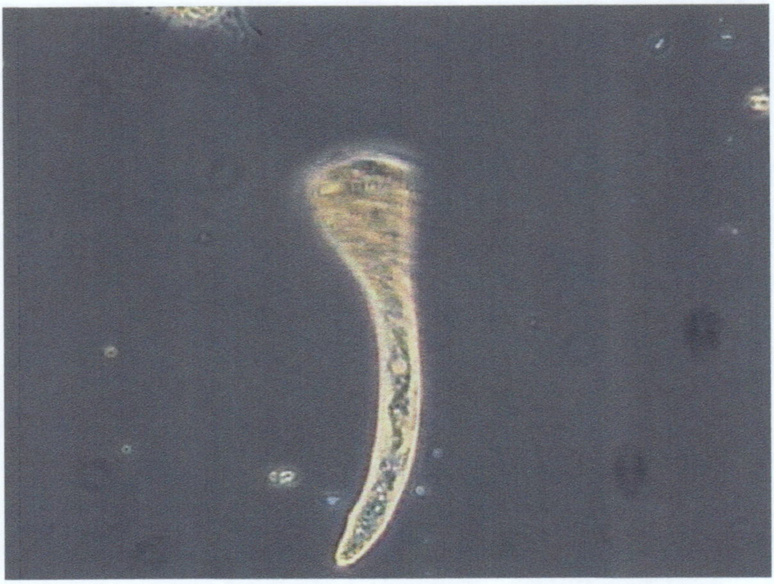

Figure 5-16 ***Stentor*** **(1000× phase contrast).**

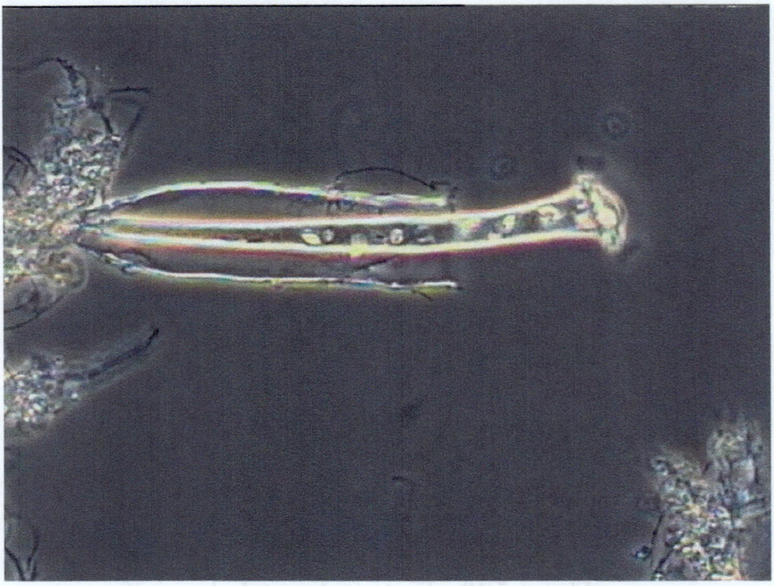

Figure 5-17 ***Vaginacola*** **(400× phase contrast).**

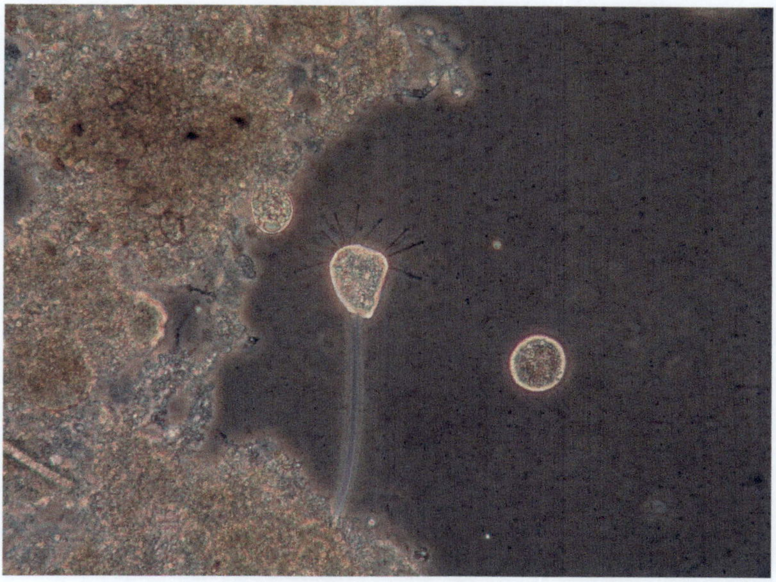

Figure 5-18 ***Suctoria*** **(400× phase contrast).**

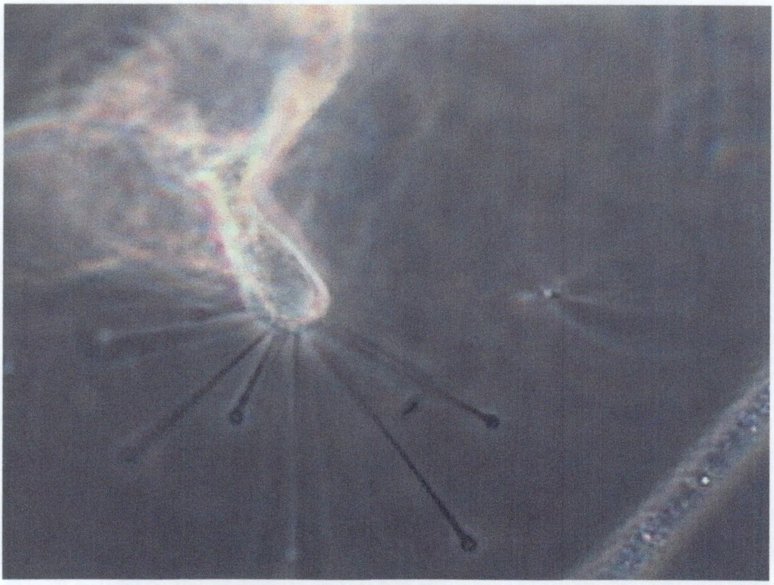

Figure 5-19 ***Suctoria,*** **long tubular tentacles (1000× phase contrast).**

Protozoa in Activated Sludge

Although protozoa play a secondary role in wastewater treatment, the relative abundance of these types of microorganisms indicates the conditions of the activated sludge. Understanding their feeding mechanisms and their competitors will help us understand their dominance in the system.

What if large numbers of amoebae are present in the activated sludge? Samples collected from the discharge end of the aeration basin should contain very few organic nutrients. If enough time was allowed in the basin, most of the nutrients should be removed. Amoebae move slowly and have difficulty competing for food when the levels of nutrients are very low. Some amoebae will always be present in the activated sludge. However, if large numbers of amoebae are present in a sample collected at the end of the aeration basin, there is still a significant amount of untreated food in the wastewater. This may indicate one or more of the following:

- A shock load of BOD has entered the treatment system—sudden increase in food allows the amoebae to compete and multiply.
- A large amount of particulates—the amoebae favor particulate food; bacteria prefer soluble nutrients.
- Low DO—because amoebae are the most primitive of the protozoa and move slowly, they require less oxygen and can compete with other protozoa when oxygen is limited.

What if flagellates dominate in the activated sludge? Flagellates prefer soluble nutrients and dead or decaying material. They compete with the bacteria for soluble food. Hence, they can only dominate when the nutrient level is quite high. Their dominance in the activated sludge treatment system indicates a high level of soluble nutrients in the wastewater. They also dominate immediately after a toxic event because they feed on dead material. Some of the larger flagellates also eat bacteria. However, they cannot compete with the ciliates for the bacteria.

Amoebae and flagellates are always present in activated sludge; however, they dominate early in the treatment process. As the bacteria begin to increase in number, soluble organic nutrients are consumed rapidly. The amoebae, as the most primitive, have no built-in mechanisms that would allow them to compete for food. Although the flagellates are better equipped to move and search for food, they cannot compete with the logarithmic growth rate of the bacteria. In a continuously fed batch process, the dominance of these microorganisms is short lived.

What if free-swimming ciliates dominate in the activated sludge? The dominance of free-swimming bacteria generally indicates an abundance of active bacteria. This means it is still early in the treatment process and the nutrients have not been depleted. If a sample collected at the end of the activated sludge process contains a dominance of free-swimming ciliates, the effluent will most likely be turbid and full of nonflocculated bacteria. Usually, very little floc is formed and the solids will not settle well. Treatment is not considered complete until most nutrients are gone and the bacteria have formed floc that will settle well in the final clarifier. Increasing the sludge age will give the bacteria enough time in the aeration basin to remove most of the organic nutrients and to begin to form floc.

What if there is an abundance of crawling ciliates in the activated sludge system? Once most of the organic nutrients have been removed, the bacteria are no longer freely dispersed, but are clumped together to form floc. Free-swimming ciliates begin to lose their advantage. Crawling ciliates, however, begin to dominate because they can find food among the floc particles. An abundance of crawling ciliates means an abundance of floc, which means most of the organic material has been removed from the wastewater. It also indicates that the sludge age of the treatment system is adequate.

What if stalked ciliates dominate in the treatment system? An abundance of stalked ciliates indicates that most of the organic nutrients have been removed. As the sludge begins to age, the dominance of stalked ciliates changes from single stalks to colonial species.

Factors That Affect Protozoa

Numerous environmental factors influence the reproduction and death of protozoa. Factors such as nutrients, DO, and competition influence the abundance and types of microorganisms in the treatment system.

Like bacteria, protozoa need oxygen. A lack of DO will severely limit the type and number of these microorganisms in the system. They can survive and reproduce in most temperatures that are common to activated sludge treatment systems. However, like most organisms, their metabolism is slower in colder temperatures. In fact, their metabolic activity doubles with every 10° rise in temperature. They grow best in ambient temperatures around 15 to 25°C.

Most municipal wastewaters contain enough nutrients to support most protozoa. Typically, protozoa depend on dead and living bacteria for food, which in turn depend on the nutrients in the wastewater. Also, many protozoa compete with one another for bacteria; some compete with bacteria for soluble nutrients. Bacteria multiply much faster than protozoa. So as long as nutrients are available, bacteria will outcompete protozoa for soluble nutrients.

Ciliated protozoa compete with each other for bacteria. Their dominance depends on whether they can consume the most bacteria. As long as the bacteria are active and freely dispersed, free-swimming ciliates will have the competitive edge. They move quickly throughout the fluid, consuming bacteria at an enormous rate. As the nutrient levels decrease and bacteria begin to form floc, crawling ciliates gain the competitive edge. They can graze along the floc and consume straggling bacteria. As nutrient levels and the number of bacteria continue to decrease, stalked ciliates gain the advantage. Other ciliates must travel to find food, but stalked ciliates can create a current that will bring the food to them. As the competition for food increases, predators such as suctorians, rotifers, nematodes, and waterbears gain the advantage by feeding off other protozoa.

Factors That Affect Protozoa

Numerous environmental factors influence the reproduction and death of protozoa. Factors such as nutrients, DO, and competition influence the abundance and types of microorganisms in the treatment system.

Like bacteria, protozoa need oxygen. A lack of DO will severely limit the type and number of these microorganisms in the system. They can survive and reproduce in most temperatures that are common to activated sludge treatment systems. However, like most organisms, their metabolism is slower in colder temperatures. In fact, their metabolic activity doubles with every 10° rise in temperature. They grow best in ambient temperatures around 15 to 25°C.

Most municipal wastewaters contain enough nutrients to support most protozoa. Typically, protozoa depend on dead and living bacteria for food, which in turn depend on the nutrients in the wastewater. Also, many protozoa compete with one another for bacteria; some compete with bacteria for soluble nutrients. Bacteria multiply much faster than protozoa. So as long as nutrients are available, bacteria will outcompete protozoa for soluble nutrients.

Ciliated protozoa compete with each other for bacteria. Their dominance depends on whether they can consume the most bacteria. As long as the bacteria are active and freely dispersed, free-swimming ciliates will have the competitive edge. They move quickly throughout the fluid, consuming bacteria at an enormous rate. As the nutrient levels decrease and bacteria begin to form floc, crawling ciliates gain the competitive edge. They can graze along the floc and consume straggling bacteria. As nutrient levels and the number of bacteria continue to decrease, stalked ciliates gain the advantage. Other ciliates must travel to find food, but stalked ciliates can create a current that will bring the food to them. As the competition for food increases, predators such as suctorians, rotifers, nematodes, and waterbears gain the advantage by feeding off other protozoa.

Chapter 6

Metazoa

Metazoa are multicellular organisms and include all animals except the protozoa. They are much more complex and range from bumblebees and spiders to fish and elephants. For the purpose of studying activated sludge, we will discuss some of the smaller microscopic metazoa and their role in the wastewater treatment system.

Metazoa are larger than most protozoa and should be observed with the phase contrast condenser and no greater than the 40× objective. For some of the larger metazoa, it may be necessary to use the 10× objective.

Metazoa have very little to do with the removal of organic material from the wastewater. Although they do eat bacteria, they also feed on algae and protozoa. A dominance of metazoa is usually found in longer age systems; namely, lagoon treatment systems. Although their contribution in the activated sludge treatment system is small, their presence does indicate treatment system conditions. Rotifers, nematodes, and tartigrades (waterbear) are the most common metazoa found in the activated sludge treatment system.

Rotifers

The principal contribution of rotifers is to clarify the effluent by removing leftover bacteria, algae, or other smaller protozoa.

Sometimes, as many as four types of rotifers may be present in the treatment system at one time.

Rotifers should never dominate in the treatment system. They reach their peak after most of the nutrients have been removed from the wastewater. The presence of dead rotifers in a fresh sample of mixed liquor is a good indicator of wastewater toxicity. They are usually the first to be affected by a toxic load.

Rotifers Under the Microscope

Rotifers come in a variety of shapes and sizes and are amazingly complex. The word *rotifer* comes from the Greek word meaning "wheel bearing animal." The wheels of the rotifer are actually a structure called the *corona,* which is covered with cilia. The corona moves constantly and brings food into the rotifer or provides a means of locomotion. Some species appear to walk; others secrete a sticky substance that enables them to attach to surfaces. A rotifer also has a unique structure called a *mastax* that has many muscles that control a set of jaws. The jaw structure grinds the food that is brought down by the action of the cilia. The female rotifer is much larger than the male, whose life span is very short and whose only purpose for living is to fertilize the female. Figures 6-1, 6-2, and 6-3 show three different rotifers—Rotaria, Epiphanes, and Lecane *(rotifer with the turtle-like shell).* Most rotifers are 200 to 500 micrometers long. The species *Rotaria neptunia* (Figure 6-1) is the most common and can stretch longer than a millimeter. Epiphanes (Figure 6-2) is a very large rotifer. It can be found in almost all aquatic habitats. Some rotifers like Lecane (Figure 6-3) have a shell.

Nematodes

A nematode has a simple structure, even though it is also multicellular. It possesses a digestive, reproductive, and nervous system. Nematodes feed on bacteria, fungi, small protozoa, and sometimes other nematodes. Some have teeth and some have a spear that they can stick into their prey. They use the spear like a straw to suck in food.

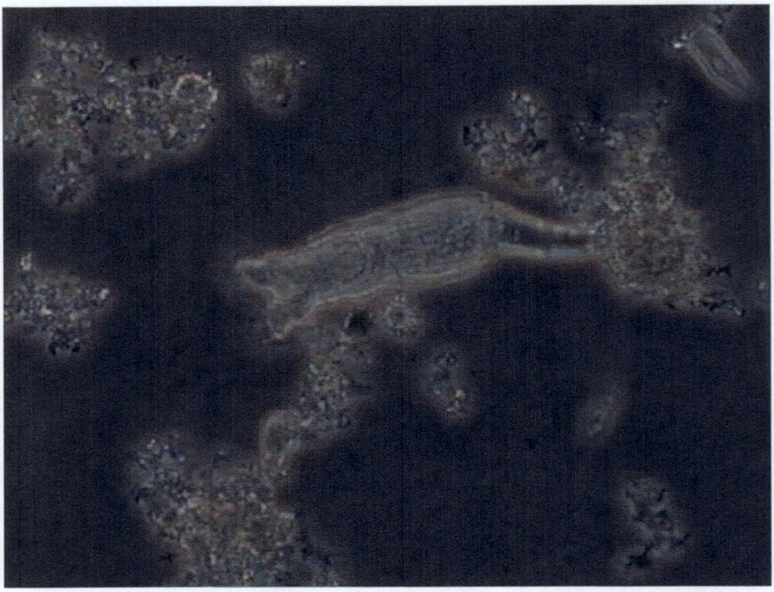

Figure 6-1 Rotifer, *Rotaria neptunia* (400× phase contrast).

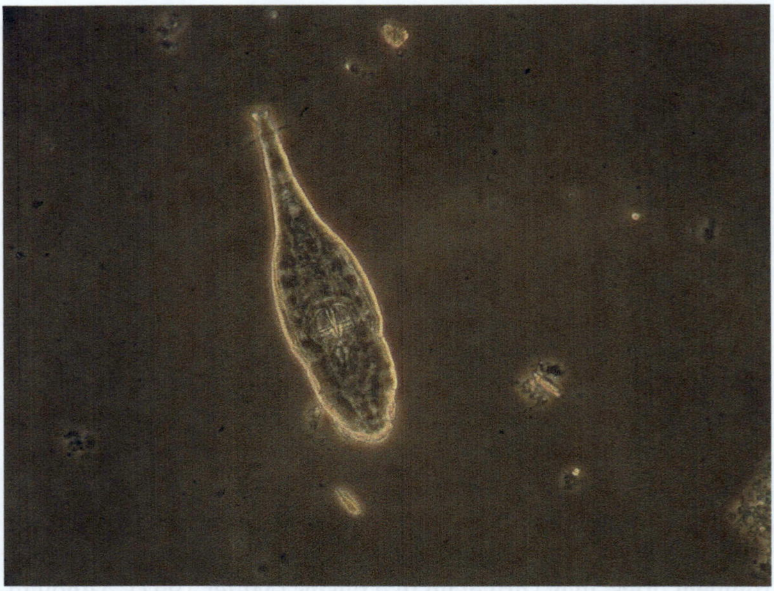

Figure 6-2 Rotifer, *Rotaria epiphanes* (400× phase contrast).

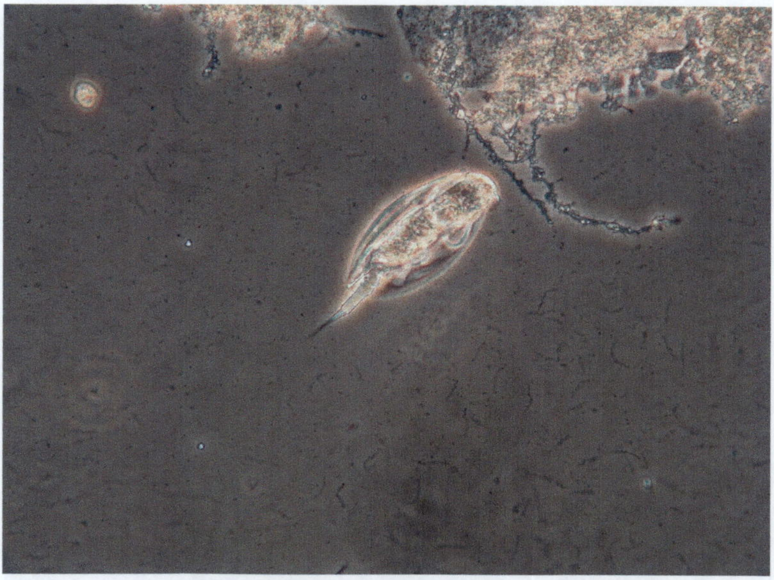

Figure 6-3 Rotifer, *Lecane,* a shelled rotifer (400× phase contrast).

Nematodes Under the Microscope

Nematodes come in various sizes. They are not segmented and move throughout the fluid by wiggling their bodies back and forth (Figure 6-4).

Tartigrades (Waterbear)

The word *tartigrade* means slow walker. Aptly called *waterbear,* this describes their slow, clumsy movement. They are aquatic microorganisms that depend on water to find food, breathe, reproduce, and move. They are commonly found in the same environment with rotifers and nematodes. They have developed ways to survive extreme environmental swings. When there is a lack of oxygen, they swell up and float for a few days. When the environment dries, they shrivel up like raisins (Mitter 1997). Although they can survive extreme environmental swings, they are sensitive

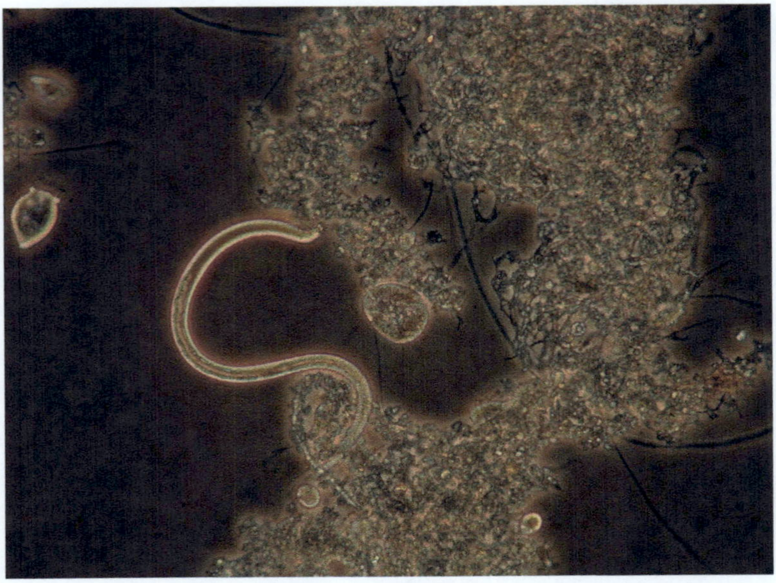

Figure 6-4 Nematode (400× phase contrast).

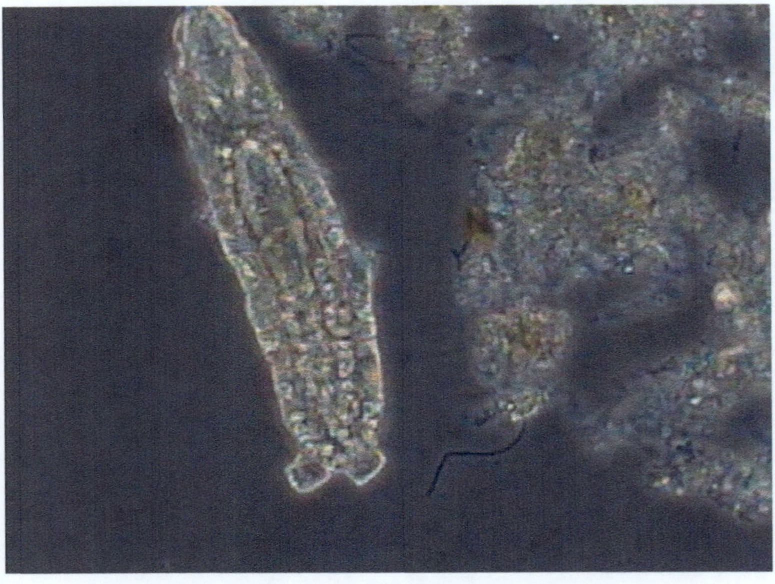

Figure 6-5 Waterbear (400× phase contrast).

to toxic conditions. Normally waterbears in the wastewater treatment system indicate that the system is efficiently removing ammonia or that other toxic substances are not present.

Waterbear Under the Microscope

Waterbears have five body segments and four pairs of short, stumpy legs with claws. They have red, orange, or green bodies. Their head has eyes with a mouth that is used to pierce their food before sucking out the inner parts. They feed on algae and small protozoa (Figure 6-5).

Reference

Mitter, W.R. 1997. Tartigrades: Bears of the Moss. In *The Kansas School Naturalist*. Empona, Kansas: Emporia State University.

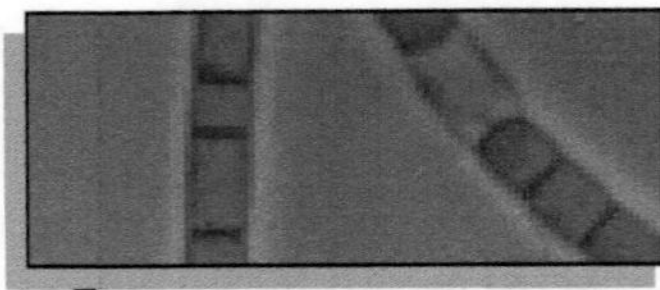

Chapter 7

Filamentous Bacteria

Under normal conditions in the activated sludge treatment system, bacteria occur singly, in pairs, in small groupings, or in short chains. However, when operational conditions drastically change, the bacteria that grow in long filaments begin to gain an advantage. Changes in temperature, pH, DO, sludge age, or even the amounts of available nutrients such as nitrogen, phosphorus oils, and grease can affect these bacteria. The dominance of filamentous bacteria in the activated sludge treatment system can cause problems with sludge settling. At times excessive numbers of filamentous microorganisms interfere with floc settling and the sludge becomes bulky. This bulking sludge settles poorly and leaves behind a turbid effluent. Some filamentous microorganisms may cause foaming in the aeration basin and clarifiers. Filamentous bacteria often live in the system in a nonfilamentous growth phase. In this phase they are harmless and contribute to the overall removal of BOD. Some filamentous bacteria are always present, even in a well-operating system. The presence of some filamentous bacteria is helpful. The filaments can provide a support structure for bacteria to attach to as they form floc. However, when they gain dominance and begin to extend beyond the floc, they may cause problems with sludge settling.

Identification

The identification of the different types of filamentous microorganisms typically involves the use of taxonomic keys. These keys lead you through a series of "yes" or "no" options that direct you to the possible causative filament. The most widely used keys are the ones prepared by Eikelboom (1977) and Jenkins, Richard, and Daigger (1993). These keys help positively identify the filaments. In fact, Eikelboom observed many of the filamentous organisms listed as types. Typically, to identify filamentous bacteria, one would look at several characteristics, which are described in the following paragraphs.

Filament shape and length. Filamentous bacteria grow in a variety of sizes and shapes. Filaments are curved, straight, or irregularly shaped. They also come in different lengths. A microscope eyepiece with a measuring grid must be used to measure the length or width of the cell or filament.

Cell shape and length. Since they are actually filaments of bacteria, the individual cells of the filament are made up of bacteria of different shapes, sizes, lengths, and widths. Some filaments are made up of round bacteria; others are rectangular, square, discoid, oval, rod, or even barrel-shaped.

Cell septa (with or without indentation). Septa are the cross walls that divide the filament into individual cells. Some filaments are made up of one long cell and septa cannot be seen. Others have septa that are not visible. Some of the cells may even indent at the septa; others do not (Figure 7-1).

Sheath. The sheath is a nonliving rigid structure that covers the filament. It resembles a clear soda straw with individual cells stacked inside (Figure 7-2).

Epiphyte (attached growth). Another characteristic that is common only to sheathed filaments is the presence of single-celled bacteria that grow perpendicular along the outer surface of the filamentous organism. This attached growth is called *epiphyte* (Figure 7-3).

Branching. Some filaments exhibit true or false branching. False branching is caused by the physical attachment of one or more branches to the main filament. There is no continuum

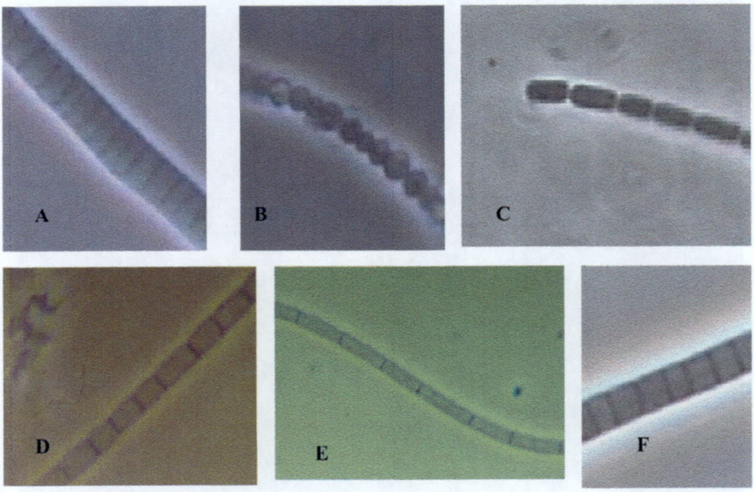

Figure 7-1 Filament cell shapes: (a) discoud, (b) round (cocci), (c) round-ended rods (bacilli), (d) and (e) rectangular (bacilli), and (f) square-shaped (bacilli). All the filaments except (e) have indentations at the cell septa.

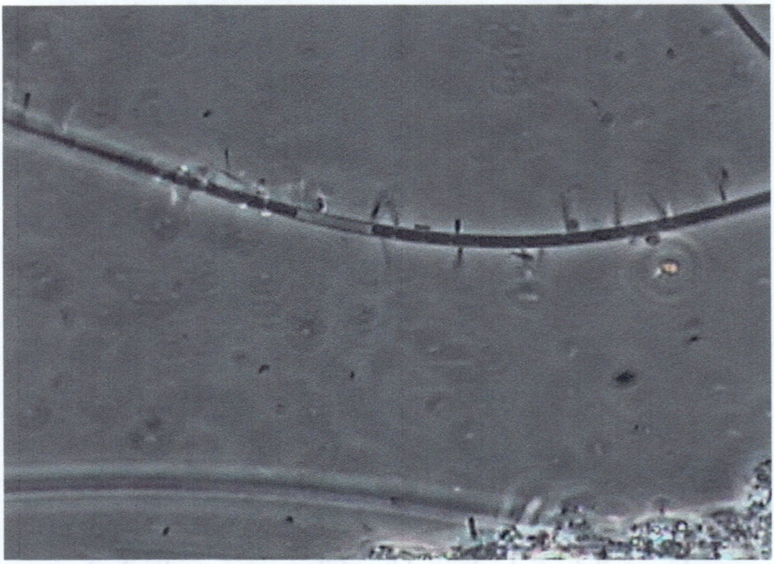

Figure 7-2 Sheath. Cells are missing (1000× phase contrast).

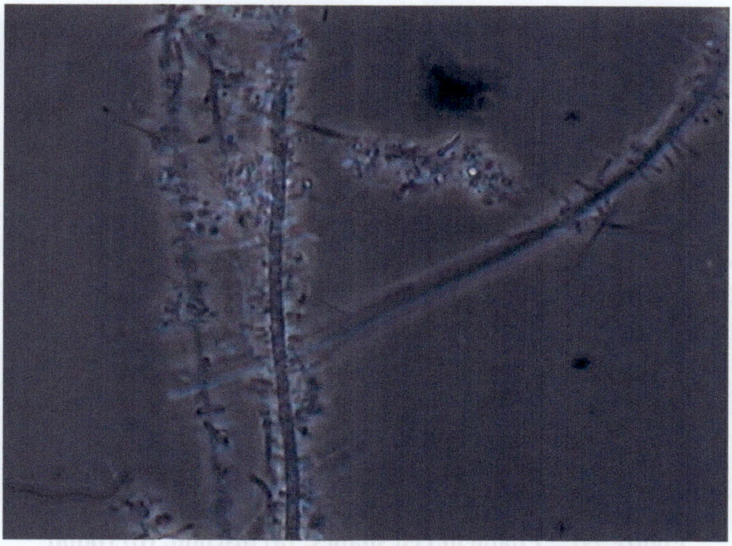

Figure 7-3 Attached growth. Some sheathed filaments are commonly seen with bacteria growing along the surface of the filament (type 0041 1000× phase contrast).

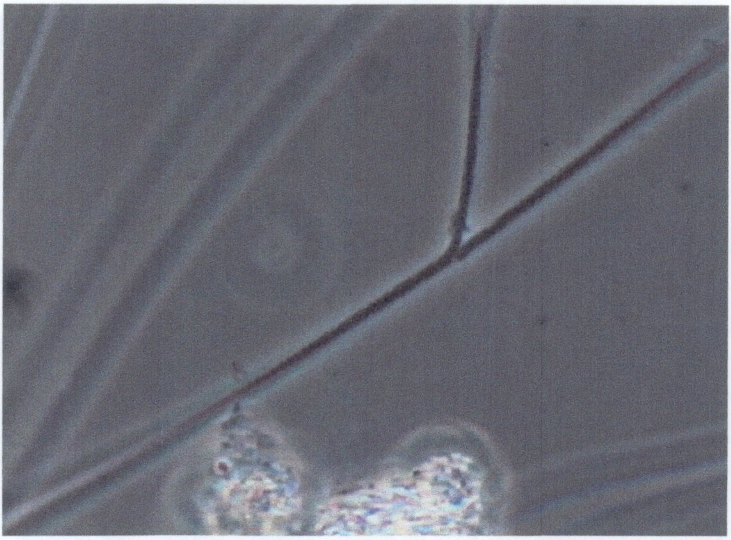

Figure 7-4 False branching. Filaments are attached with no continuum (*S. natans* 1000× phase contrast).

between the branches (Figure 7-4). True branching is actually the result of the emergence of new cells with continuum between branches (Figure 7-5).

Sulfur granules. Some filaments can accumulate sulfur granules inside their cells. When viewed under the microscope with phase contrast, the granules look like light spots within the cells of the filament (Figure 7-6).

Motility. Motility, the filament's ability to swim, is probably the easiest characteristic to identify.

Looking for all these characteristics and measuring lengths and widths of cells and filaments can be very time consuming and requires a lot of skill. Let's look at a way to make this identification process a bit easier.

Simplified Identification

Filamentous bacteria can contribute to sludge settling problems by causing sludge to bulk or they can cause foaming in aeration basin and clarifiers. First, the dominance of filamentous bacteria

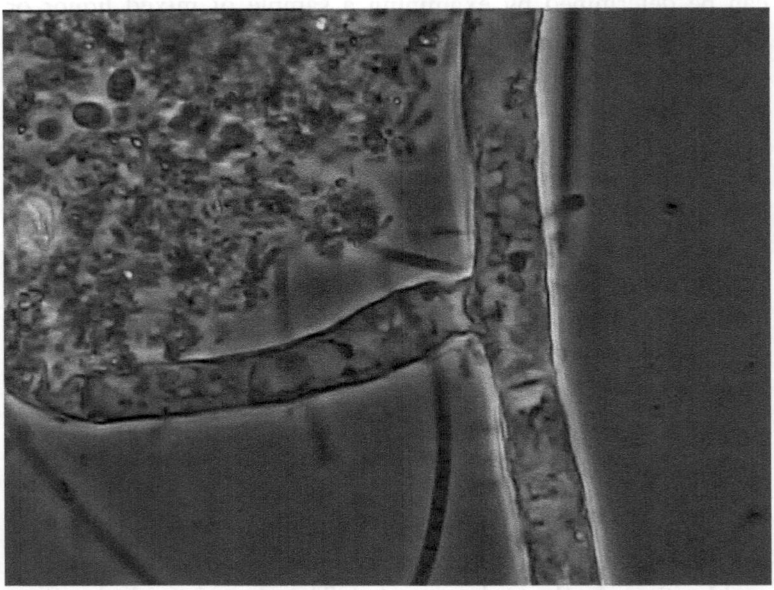

Figure 7-5 True branching. Fungus (1000× phase contrast).

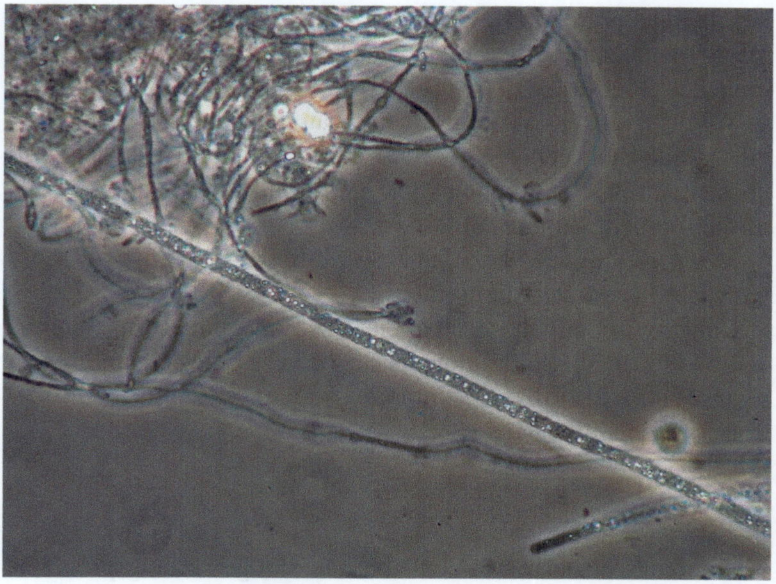

Figure 7-6 Sulfur granules in *Thiothrix* (1000× phase contrast).

as the cause of the bulking or foaming must be confirmed. This can be determined by examining a sample of mixed liquor or foam under the microscope with the 20× or 40× objective. If the filamentous bacteria are causing the sludge to bulk, their dominance will be apparent in the sample. If the sludge is bulking, examine the mixed liquor. Numerous filaments will be in the floc and extend into the bulk fluid. If the system is foaming, examine a sample of the foam. If filamentous bacteria are causing the foaming, the filaments will be dominant in the foam.

Once it has been established that the bulking or foaming is due to the excess growth of filamentous bacteria, you can begin the process of identifying certain characteristics. Begin by separating the filaments into two categories: foaming filaments and bulking filaments.

Foaming Filaments

Filaments that cause foaming have hydrophobic cell walls. The cells are covered with a waxy coating that resists water and enables them to float when aerated. They have very little effect on the settleablility of the sludge or the overall treatment of the

wastewater, unless they cause excessive amounts of foam. However, aesthetically they cause the treatment system to appear to be running poorly.

Only three filaments, *Nocardia* spp., *Microthrix parvicella*, and type 1863, are the major causes of foaming in activated sludge treatment systems. These are very easy to identify. On rare occasions, foaming has been associated with filament type 0092. It is not known whether type 0092 causes foaming or whether it is often present under the same conditions as *Microthrix* and is therefore trapped in the foam.

To identify the type of filament causing the foam, first collect a sample of foam and make a smear. Allow the smear to air dry and then stain it with Gram stain. Gram stain will help to narrow your choices to two: gram-positive or gram-negative. Two of the filaments are gram-positive and the other is gram-negative. Observe the gram-stained slide under a drop of immersion oil with the 100× objective. Purple-stained filaments are gram-positive and pink-stained filaments are gram-negative. Gram-positive filaments are either *Nocardia* or *M. parvicella*. A gram-negative filament is probably type 1863 (Figure 7-7).

Type 1863 looks like a pink-dashed line and is easy to identify (Figure 7-8). *Nocardia* and *M. parvicella* also have a unique appearance, even though they are both gram-positive. *M. parvicella* looks like purple spaghetti and is the one foaming filament that is often associated with both bulking and foaming. *Nocardia*,

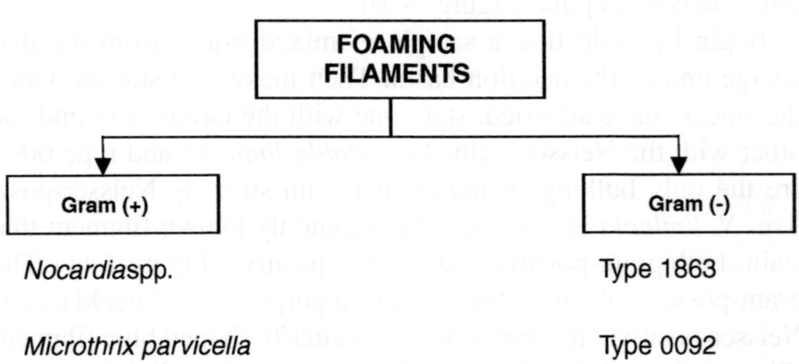

Figure 7-7 Foaming filaments.

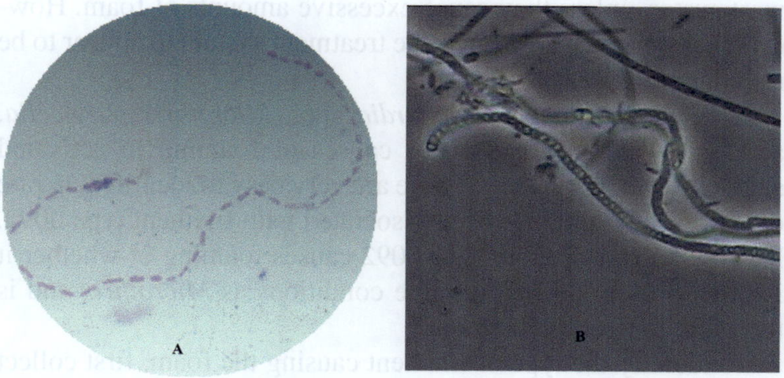

Figure 7-8 Foaming filament type 1863: (a) gram-negative (1000× brightfield); (b) wet mount (1000× phase contrast).

on the other hand, looks like small patches of irregularly branched short filaments (Figure 7-9).

Bulking Filaments

Foaming filaments are easy to identify by doing a Gram stain. The process of identifying bulking filaments, on the other hand, requires more investigating. The Gram or Neisser stain is helpful but provides only limited information. Most bulking filaments are gram-negative and Neisser-negative. A few filaments are gram-variable, with both gram-positive and gram-negative characteristics. This means that parts of the filament stain purple and other parts stain pink (Figure 7-10).

Begin by collecting a sample of mixed liquor from the discharge end of the aeration basin. Then make two smears. Once the smears have air dried, stain one with the Gram stain and the other with the Neisser stain. *Nostocoida limicola* and type 0092 are the only bulking filaments that stain strongly Neisser-positive. *N. limicola* is also the only commonly known filament that stains both gram-positive and Neisser-positive (Figure 7-11). The gram-positive *N. limicola* looks like a purple beaded necklace; as Neisser-positive, it appears as an irregularly shaped blue filament (Figures 7-12, 7-13). Type 0092 looks like blue, slightly curved sticks (Figure 7-14).

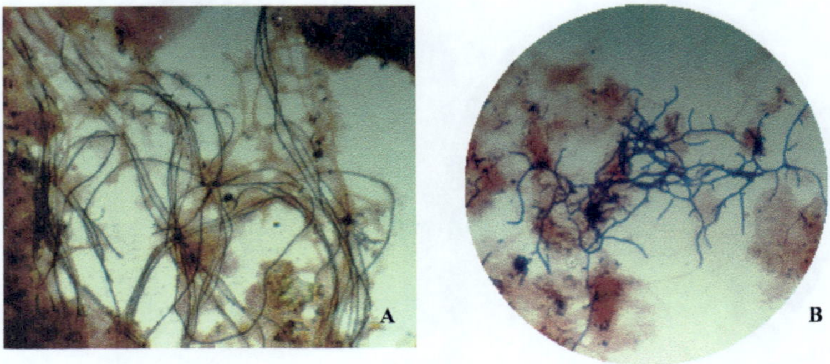

Figure 7-9 Foaming filaments: (a) *Microthrix parvicella* looks like purple spaghetti; (b) *Nocardia* spp. with short branched filaments.

If you're lucky, and *N. limicola* or type 0092 is the filament that causes bulking problems, you won't have to look any further. There are, however, three filaments that stain gram-variable. If the filament is gram-variable, you have narrowed your choices to

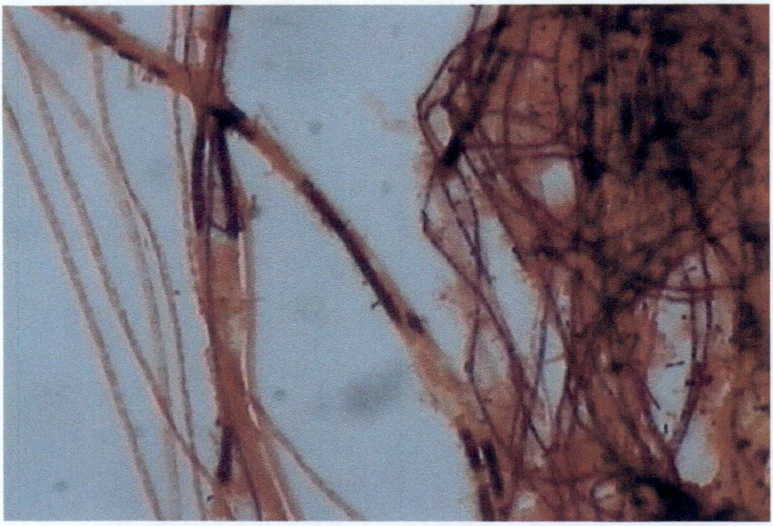

Figure 7-10 Bulking filaments gram-negative *Thiothrix* shown at the far left; gram-variable type 0041 shown in the middle staining gram-variable; gram-positive *Microthrix* shown at the far right.

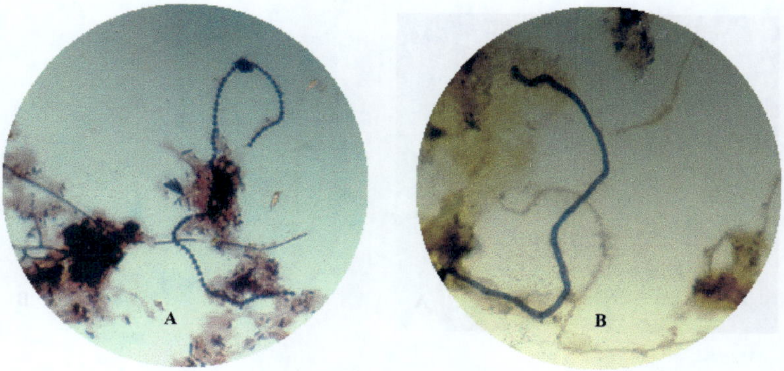

Figure 7-11 *Nostocoida limicola* (a) gram-positive, (b) Neisser-positive. This is the only known filament that is both gram-positive and Neisser-positive.

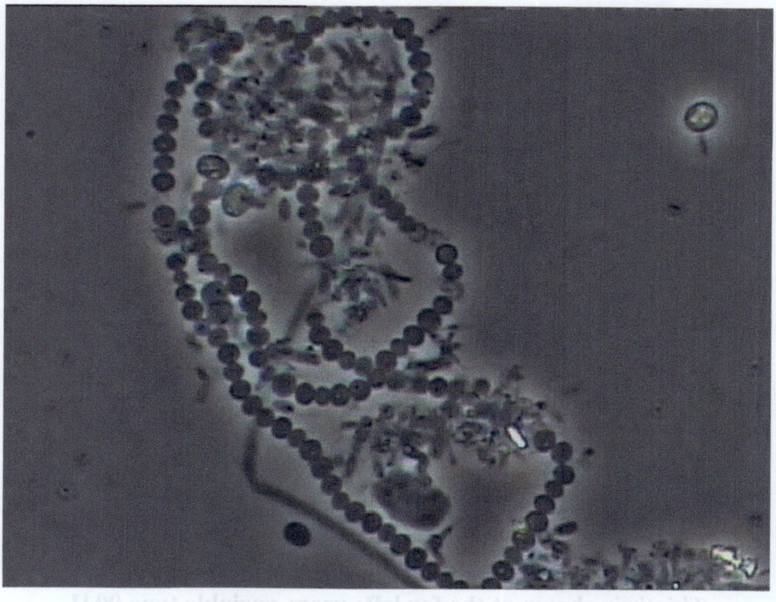

Figure 7-12 *Nostocoida limicola* type 2 (1000× phase contrast).

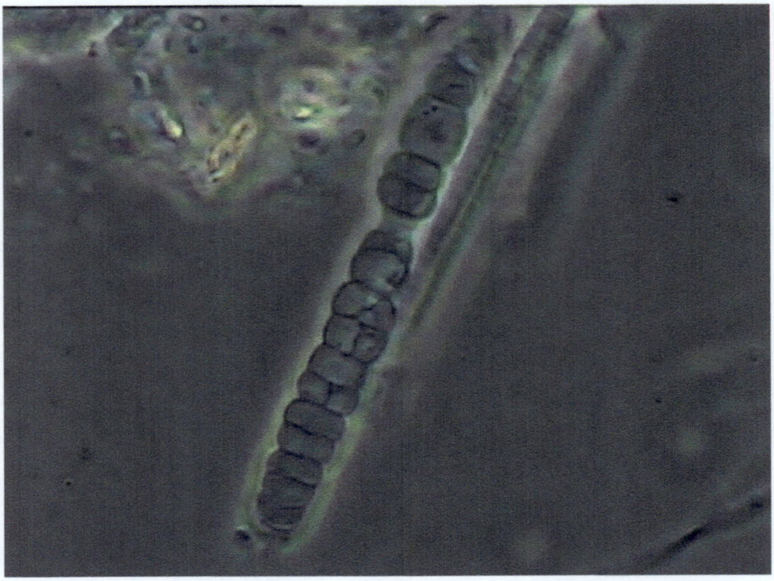

Figure 7-13 *Nostocoida limicola* type 3 (1000× phase contrast).

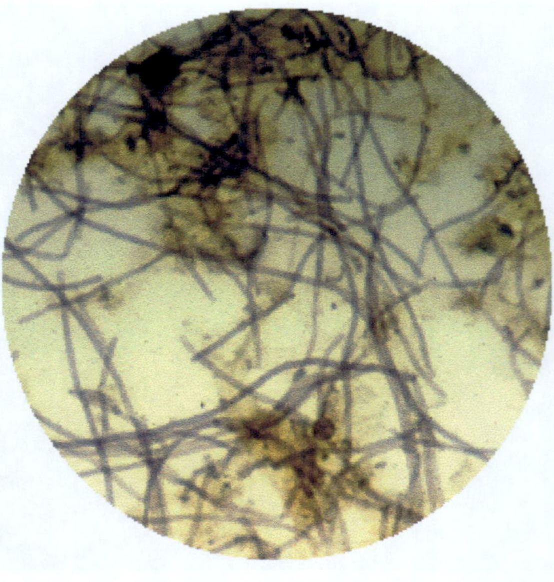

Figure 7-14 Neisser stain. Filament type 0092 stains Neisser-positive (1000× brightfield).

three, but the rest of the bulking filaments are all gram-negative (Figure 7-15).

Filament types 0041, 0675, and 1851 are all gram-variable. They can be distinguished quite easily. Types 0041 and 0675 look alike. They both have square-shaped cells, but type 0041 is larger. Type 1851 is quite different. It is a thin filament and it grows in bundles (Figures 7-16, 7-17, 7-18).

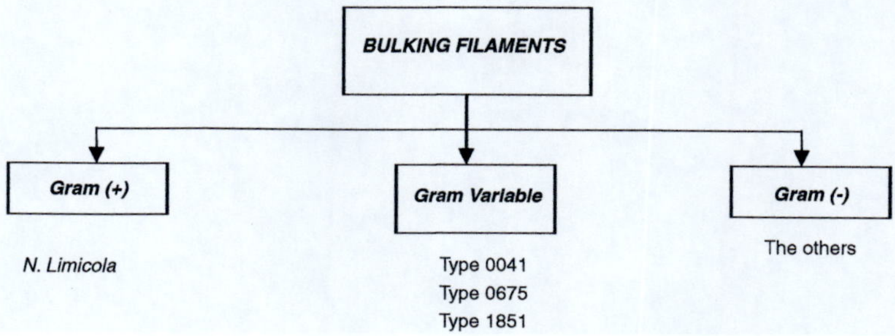

Figure 7-15 Bulking filaments

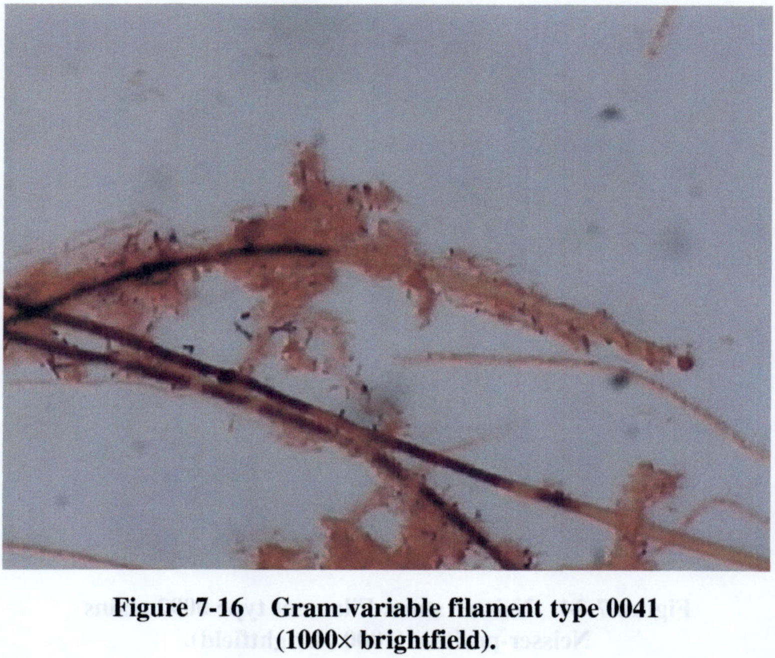

Figure 7-16 Gram-variable filament type 0041 (1000× brightfield).

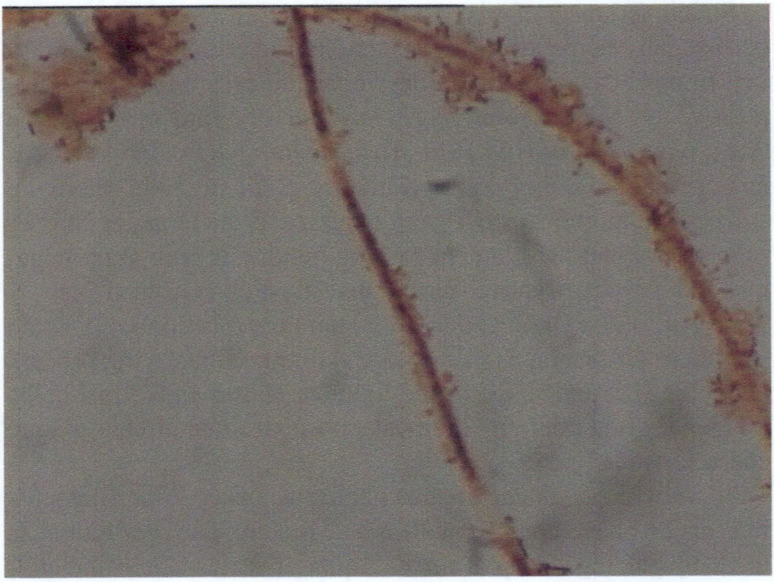

Figure 7-17 Gram-variable filament type 0675 (1000× brightfield).

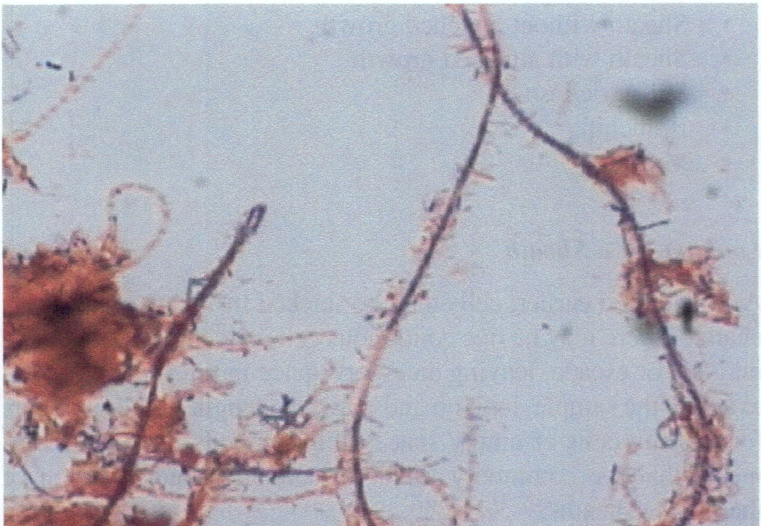

Figure 7-18 Gram-variable filament type 1851 (1000× brightfield).

Staining a smear of mixed liquor with the Gram or Neisser stain occasionally helps identify filamentous bacteria that cause bulking. In most cases, staining does very little except identify gram-positive or Neisser-positive *N. limicola* or gram-variable type 0041, type 1675, or type 1851. This narrows the field to a few filaments that are either gram-positive or gram-variable. There are still several more that stain gram-negative. If the Gram or Neisser stain shows that the filament causing bulking is both gram-negative and Neisser-negative, further investigation is needed.

As part of the Gram and Neisser staining procedures, allow the sample to air dry on a slide before staining. This drying process alters the size and shape of the filament and its individual cells. To identify other bulking filaments, observe some of their unique characteristics.

To view these characteristics unaltered, observe the filaments live under phase contrast with a wet mount. The wet mount is made by placing a drop of mixed liquor on a clean slide and covering it with a cover slip. If you view the slide with the phase contrast condenser under oil with the 100× objective, you may recognize one or more of the following characteristics:

- Sheath without attached growth
- Sheath with attached growth
- Sulfur deposits
- Branching
- Motility

Looking for a Sheath

As mentioned earlier, cells may be stacked inside a sheath like a chain or there may be one continuous cell inside the sheath. Cells may die or escape, leaving an empty space in the sheath. As you examine the sample, look up and down the length of the filaments for missing cells or empty spaces between cells. Of all the filaments that are commonly found in activated sludge, seven of them have sheaths:

- Type 1701
- Type 0041

- Type 0675
- Type 1851
- *Sphaerotilus natans*
- *Thiothrix* spp. (types 1 and 2)
- *Haliscomenobacter hydrossis*

Looking for a Sheath With Attached Growth

Some sheathed filaments have attached growths and resemble bottle brushes. Of the common sheathed filaments, four are often seen with attached growths. The sheathed filaments that have a type and number name are commonly seen with attached growths. This makes it easier to remember. These filaments can also grow in the treatment system without the attached growths. This usually means that conditions are such that they can grow quite rapidly. This also makes them a little more difficult to identify (Figure 7-19).

Identifying the sheath will help to narrow your options to seven, but additional observations are necessary to determine which dominates in the treatment system. Types 1701, 0041, and 0675 are often seen with heavy attached growths. However, the growth is only sparsely attached to type 1851 (Figures 7-20, 7-21). Remember that type 1851 is gram-variable and grows in bundles. So, even if you did not Gram stain the sample, if you see a filament with sparse attached growth growing in bundles, it is

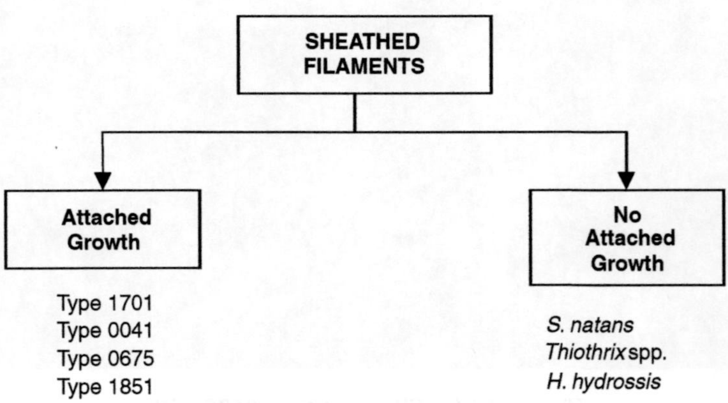

Figure 7-19 Sheathed filaments.

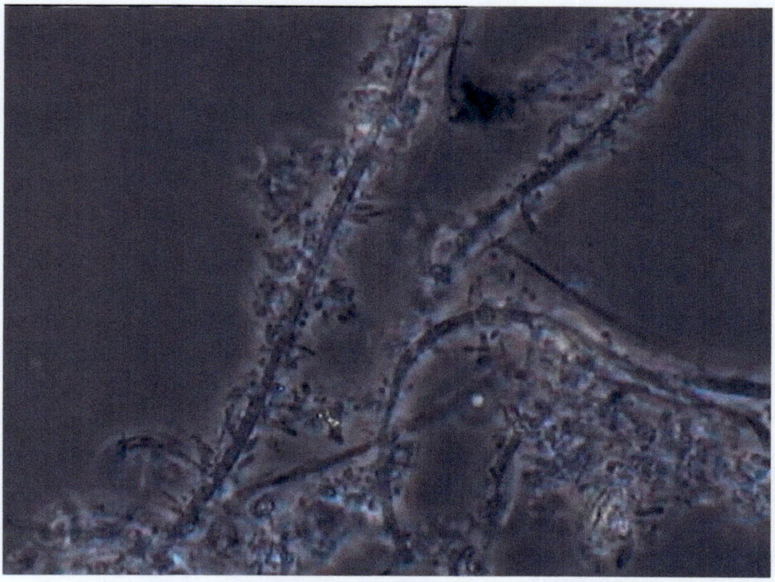

Figure 7-20 Filament type 0041 with heavy attached growth (1000× phase contrast).

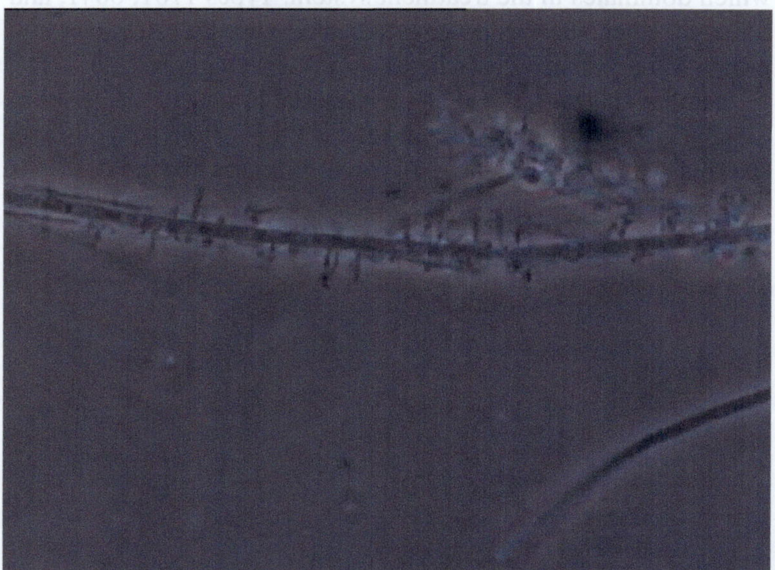

Figure 7-21 Filament type 0675 with attached growth (1000× phase contrast).

probably type 1851 (Figure 7-22). Types 0041 and 0675 are also gram-variable and, as described earlier, type 0041 is larger than type 0675 and both have square-shaped cells.

The size and shape of the individual cells will also help to identify the filaments. Type 1701 has cells that look like round-ended rods or link sausages that are stacked in a tight-fitting sheath. Type 1701 is a very thin filament. Even if it has heavy attached growth, it is easy to identify (Figure 7-23; Table 7-1).

Looking for a Sheath Without Attached Growth

S. natans, *Thiothrix* spp., and *H. hydrossis* are sheathed filaments but are generally not seen with attached growth. Nevertheless, they are very different from each other and easy to identify. *S. natans* also has sausage-shaped cells like type 1701, but it is more than twice the width of type 1701. It is also the only filament the exhibits false-branching (Figure 7-24). *Thiothrix* types 1 and 2 have barrel-shaped cells. However, type 1 is about twice

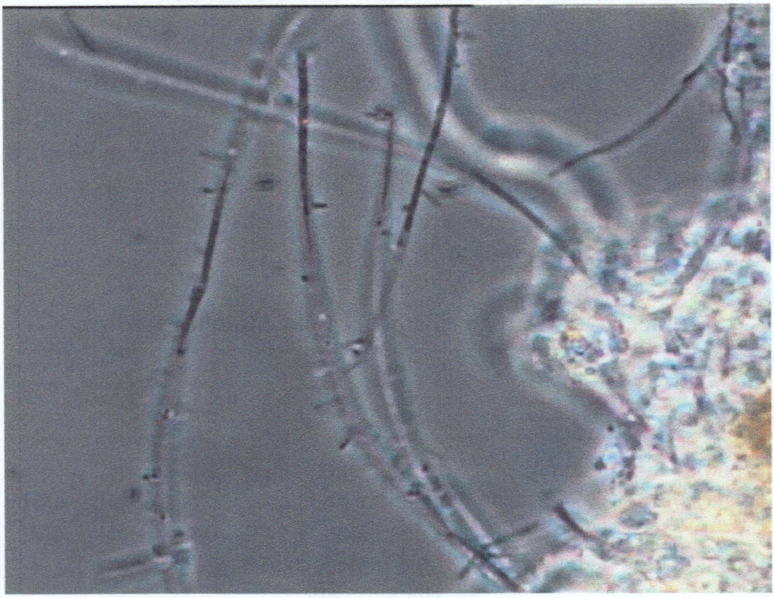

Figure 7-22 Filament type 1851 with sparse attached growth (1000× phase contrast).

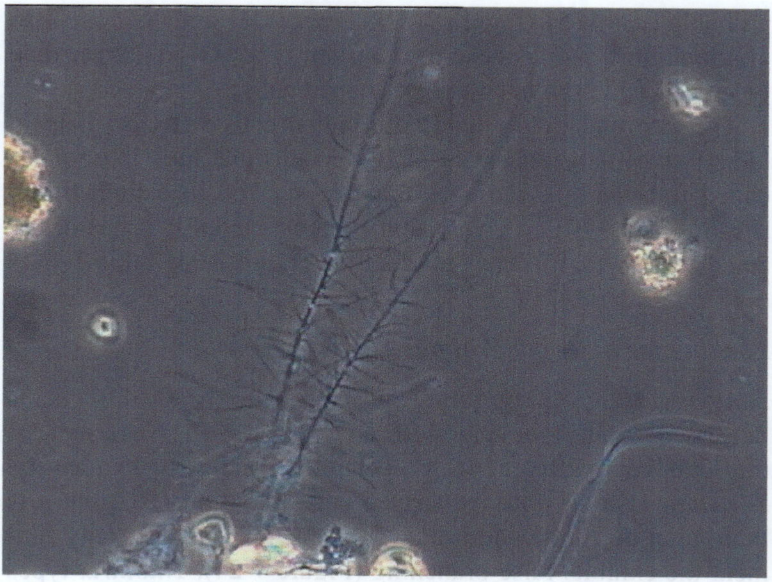

Figure 7-23 Filament type 1701 with heavy attached growth (1000× phase contrast).

the width of type 2. Another characteristic that distinguishes *Thiothrix* from *S. natans* and *H. hydrossis* is the storage of sulfur granules within the cells (Figure 7-25). *H. hydrossis* is such a straight, thin filament that it is difficult to detect a sheath. *H. hydrossis* most commonly protrudes from the flock like thin pins in a cushion (Figure 7-26; Table 7-2).

Table 7-1 Sheathed Filaments With Attached Growth

Filament Name	*Attached Growth*	*Other Characteristics*
Type 0041	Heavy	Square-shaped cells
Type 0675	Heavy	Square-shaped cells; looks like type 0041 but smaller
Type 1701	Heavy	Round-ended, rod-shaped (sausage-shaped) cells
Type 1851	Sparse	Rectangular-shaped cells; grows in bundles

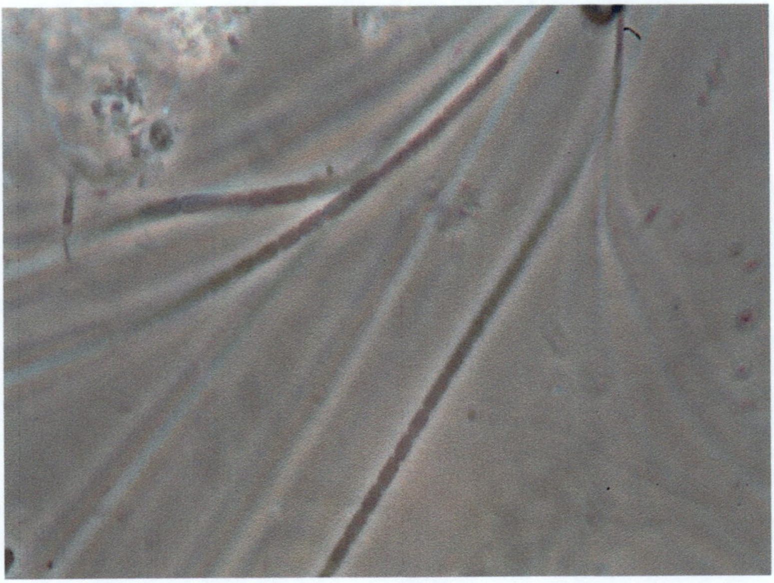

Figure 7-24 ***Sphaerotilus natans*** **with its characteristic sausage-shaped cells (1000× phase contrast).**

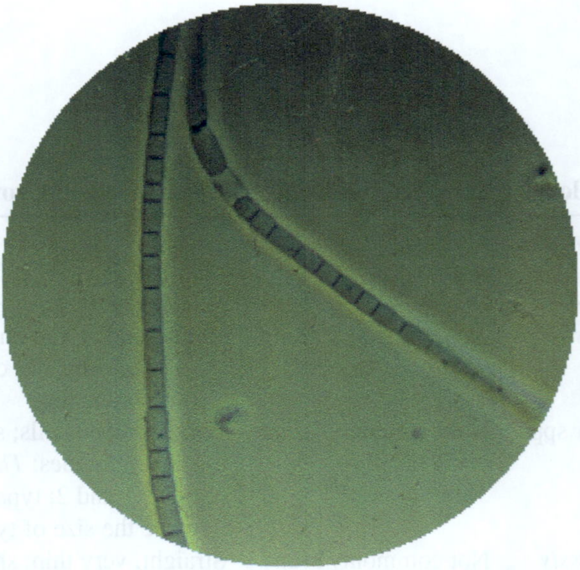

Figure 7-25 ***Thiothrix*** **spp. (1000× phase contrast).**

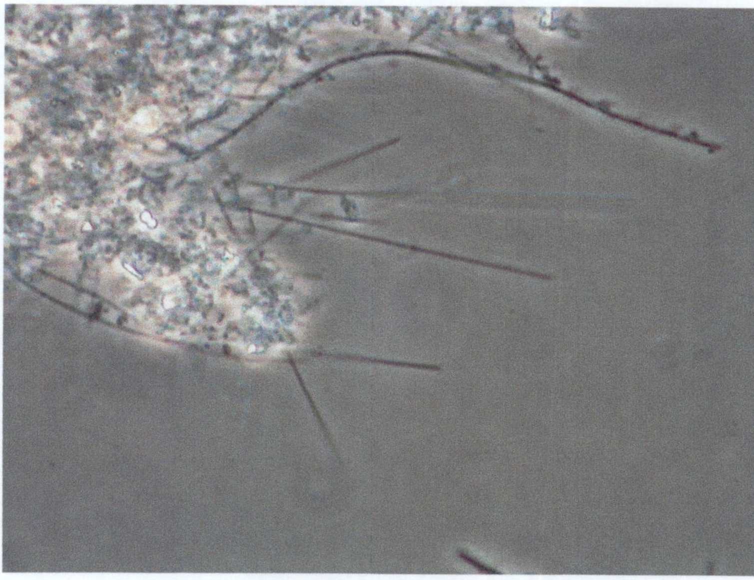

Figure 7-26 *Haliscomenobacter hydrossis,* a thin straight filament.

Table 7-2 Sheathed Filaments Without Attached Growth

Filament Name	*Attached Growth*	*Other Characteristics*
S. natans	Not commonly seen	Round-ended, rod-shaped (sausage-shaped) cells; false branching
Thiothrix spp.	Not commonly seen	Barrel-shaped cells; stores sulfur granules; *Thiothrix* types 1 and 2; type 1 is twice the size of type 2
H. hydrossis	Not commonly seen	Straight, very thin; sheath difficult to detect

Looking for Sulfur Granules

If the bulking filament at the treatment plant is gram-negative, Neisser-negative, or does not have a sheath (with or without attached growth), there are other unique characteristics to look for. Under phase contrast, look for sulfur granules within the cell. This can often be observed in systems where reduced sulfur compounds such as hydrogen sulfide are present. Reduced sulfur compounds are found in septic wastewater. Four filament types, type 021N, *Thiothrix* spp., type 0914, and *Beggiatoa*, are commonly seen with intracellular sulfur granules.

Type 021N is an unusual filament. The cells are discoid shaped and look like short, fat barrels or hockey pucks stacked on top of each other (Figure 7-27). Sulfur granules can be seen within the cells. The cells of type 021N stain gram-negative but display a slight reaction to the Neisser stain by turning a soft gray color.

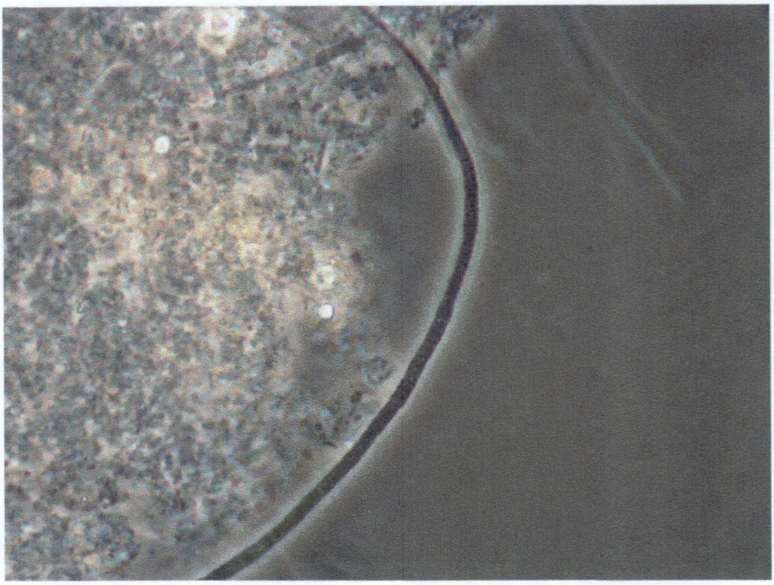

Figure 7-27 Filament type 021N with intracellular sulfur granules (1000× phase contrast).

The cells of *Thiothrix* types 1 and 2 are barrel shaped. Sulfur granules are more readily seen in the smaller type 2 (Figures 7-28, 7-29). *Thiothrix* 1 is relatively large and the sulfur granules appear much smaller.

Type 0914 is smaller but can be easily distinguished by its rectangular-shaped cells and rectangular-shaped sulfur granules. *Beggiatoa* is motile; it glides slowly in the surrounding fluid (Figure 7-30; Table 7-3).

Looking for Branching

Although branching is easy to identify, only two of the bacterial filaments common to activated sludge exhibit branching. One, *S. natans*, shows evidence of false branching (Figure 7-31). *Nocardia* exhibits true branching. Fungus is less common in activated sludge, but also exhibits true branching.

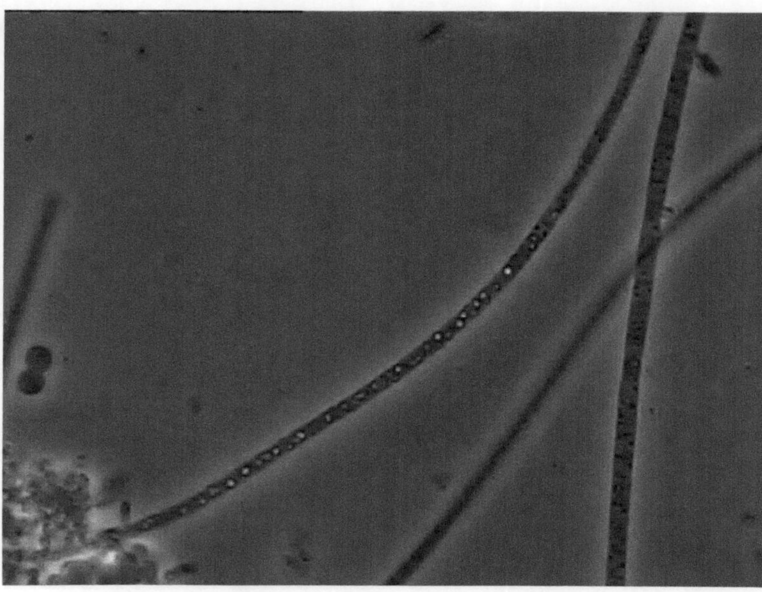

Figure 7-28 *Thiothrix* type 1 with sulfur granules (1000× phase contrast).

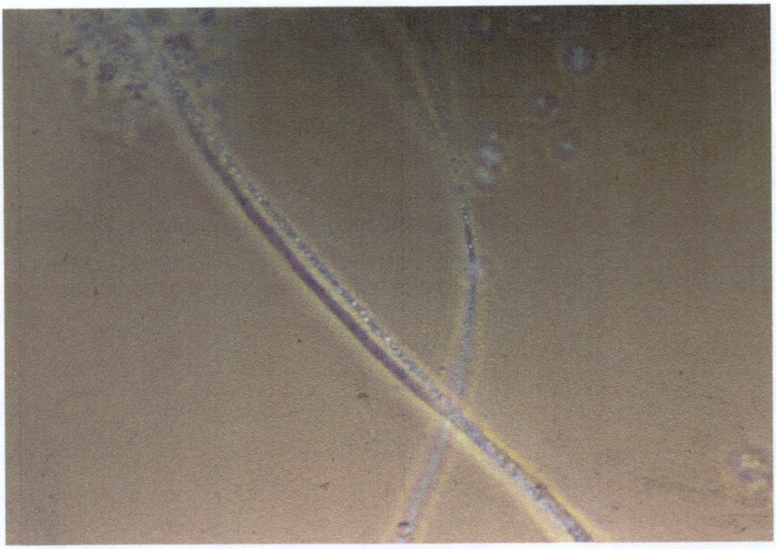

Figure 7-29 ***Thiothrix*** **type 2 with sulfur granules (1000× phase contrast).**

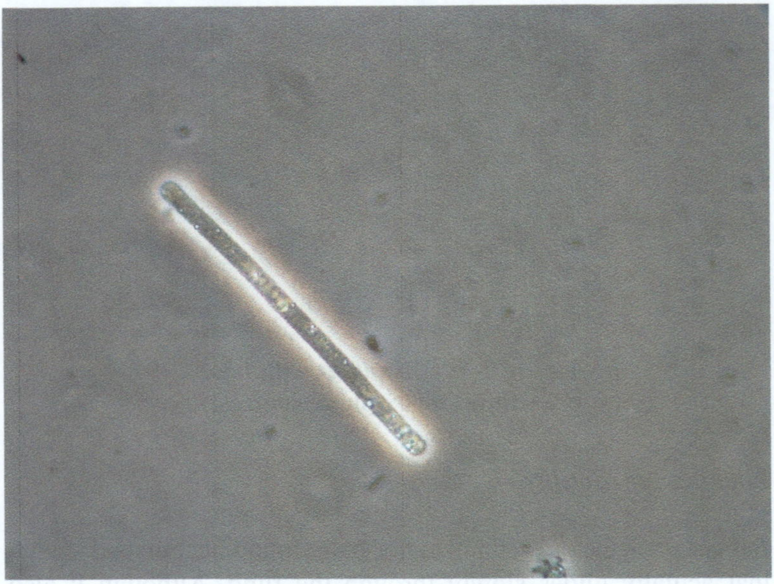

Figure 7-30 ***Beggiatoa*** **with sulfur granules. This filament glides through the fluid. It is the only motile filament.**

Table 7-3 Filaments With Sulfur Granules

Filament Name	*Other Characteristics*
Type 021N	Discoid-shaped cells; look like stacked hockey pucks; round sulfur granules; slight reaction to Neisser stain (gray color)
Thiothrix spp.	Barrel-shaped cells; stores sulfur granules; *Thiothrix* types 1 and 2; type 1 is twice the size of type 2
Type 0914	Rectangular cells with rectangular-shaped sulfur granules
Beggiatoa	Motile, slowly gliding

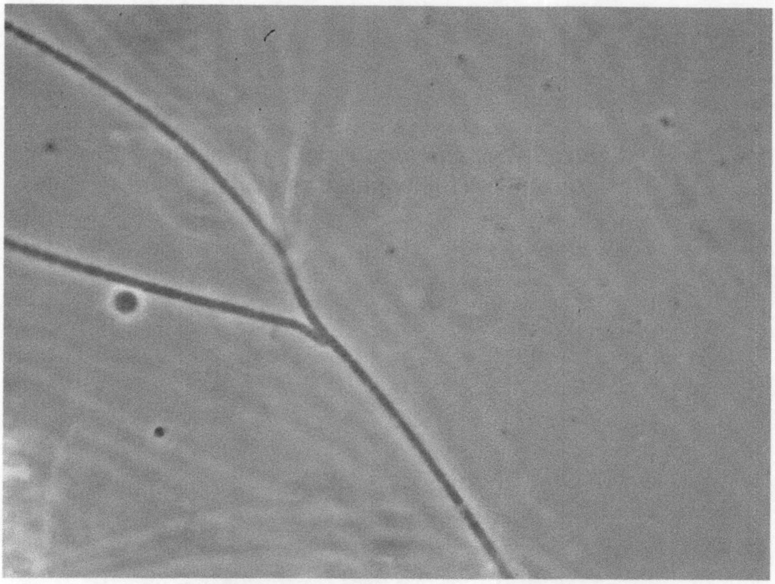

Figure 7-31 *S. natans* with false branching (1000× phase contrast).

Looking for Motility

Beggiatoa is the only filament normally found in activated sludge that actually glides slowly through the fluid. Although it also stores sulfur granules in its cells, it can be easily spotted as it swims.

Keeping It Simple

Some filaments cause foaming and some cause bulking. Identifying foaming filaments is relatively straightforward because only three normally cause foaming. Two are gram-positive and the other is gram-negative. The two gram-positive filaments are structured so differently that it is easy to tell them apart. The same is true for the gram-negative filament that causes foaming. Bulking filaments, on the other hand, require a few more steps. Observing the filaments live with the phase contrast condenser and the oil immersion lens, and grouping the filaments based on certain characteristics, will help to narrow the field. Seven of the common filaments have sheaths. Four filaments are often seen with sulfur granules within the cell, two branch, and one is motile.

Filament Couples

Some filamentous bacteria can be paired as couples. These pairs look alike and are often seen thriving under the same conditions. *S. natans* and type 1701 form one such couple. Both have sausage-shaped cells in a tight-fitting sheath and generally thrive in low DO conditions. The difference is that *S. natans* is larger and often exhibits false branching. Type 1701 is thin and is often seen with attached growth. However, if you encounter either and you are not sure which is which, it doesn't matter because they both are associated with low DO conditions.

Another filament couple is type 0041 and type 0675. Both thrive in low F/M and nutrient-deficient conditions. Both are gram-variable, have sheaths, have square-shaped cells, and are often seen with heavy attached growths. The only difference is that type 0041 is larger than type 0675. Should the widths of the filaments be measured to determine which it is? No. It really doesn't matter, since both thrive under the same conditions.

Bulking Filaments

Once filamentous microorganisms have been identified, determine the conditions that may be responsible for their dominance in the treatment system. A great deal of research has been conducted on

the causes of filamentous bulking and foaming. The major conditions that have been associated with bulking and foaming are:

- Low DO
- Low F/M ratio
- Longer sludge ages
- Septic wastewater
- Nutrient-deficient incoming wastewater
- Changes in pH
- Excess greases, oils, and fats

Jenkins, Richard, and Daigger (1993) associated filamentous bacteria that cause bulking with the following plant operating conditions and wastewater characteristics:

- Types 0041, 0675, 0581, 0961, 0803, 0092, and *Microthrix parvicella* are all associated with low F/M conditions.
- Types 1701 and *S. natans* thrive when the aeration basin DO is too low for the applied organic load.
- *Thiothrix* spp. and type 021N are favored in septic conditions. They can store sulfur granules as an energy source.
- *Thiothrix* spp. and types 021N, 0041, and 0675 are favored in nutrient-deficient wastewaters.
- Fungi are favored when the aeration basin pH level drops below 6.5.

Low F/M results when there are too many microorganisms and too little food. Low F/M conditions can be adjusted by reducing the detention time in the aeration basin. Increasing the recirculation of return activated sludge will reduce the detention time in the aeration basin, lower the MLSS, and thus increase the F/M ratio.

If the filamentous microorganisms are associated with low DO, the operator can simply adjust the aeration rate to meet the demand in the aeration basin. Sometimes the DO level in the aeration basin is sufficient. In this case, look for areas of the plant or the sewer system where there is oxygen stress. These areas can be breeding grounds for filaments favored by low DO conditions. For example, oxygen stress can occur in equalization tanks

that are not aerated before discharge into the aeration basin or in aeration basins where there is incomplete mixing. Maintaining high sludge blankets in clarifiers can also cause low DO in the system.

Increasing the DO in the aeration basin will also eliminate septicity. However, for septic wastes that enter the treatment system, pre-aeration may be necessary. Septicity can result when incoming wastewater sits stagnant in pipes, waiting to be pumped into the treatment system.

Nutrient-deficient wastewater can be supplemented with the deficient nutrient. Determine the BOD:N:P ratio at the point where the wastewater enters the aeration basin. This will give a better measurement of the amount of available nutrients for the microorganisms. Determine the amount of nitrogen available by analyzing for ammonia nitrogen and determine the amount of available phosphorus by measuring ortho-phosphorus. Urea can supplement ammonia and phosphoric acid can supplement phosphorus.

The wastewater pH can be adjusted by adding a caustic solution to increase the pH.

Foaming Filaments

The foaming filaments generally are associated with excess greases, oils, and fats. *M. parvicella* and *Nocardia* are the most common causes of activated sludge foaming. Types 1863 and 0092 have also been known to cause foaming. Richard (1989) suggests that *M. parvicella* is associated with low F/M operation, high wastewater grease and fat content, and colder temperatures. Richard also suggests that growth of *M. parvicella* is encouraged by the presence of animal and vegetable fats. Types 1863 and 0092 are associated with low F/M operations.

About **Nocardia** *spp.*

Nocardia is the most common foam-causing filament in activated sludge. The strongly hydrophobic branches enhance sludge flotation by forming a net, enabling them to trap oil droplets and gas bubbles. *Nocardia* likely produces surface-active substances in abundance, altering the surface tension and causing foaming when

aerated. Under certain conditions, *Nocardia* produces a lipid material, which is excreted into the mixed liquor, and collects on the surface of air bubbles. This causes the bubbles and *Nocardia* to mesh together, giving the foam the appearance of chocolate parfait.

Nocardia is one of the most difficult filaments to eliminate from the treatment system. It competes successfully for several reasons. The hydrophobic cell walls and branched hyphae allow oil droplets and gas bubbles with accumulated nutrients on their surface to stick to the *Nocardia* cell walls. Also, fats, proteins, and other nutrients that float on the surface of the water accumulate and supply more nutrients than those in the bulk water. *Nocardia* can also live on dead cells and can use organic compounds adsorbed on floc that are only slowly degraded by other organisms. This allows them to survive periods when dissolved nutrients are scarce. They can also store polyphosphates as reserve material, which enables them to increase in the system even when they are virtually cut off from a nutrient supply. As a former soil inhabitant, *Nocardia* is better adapted than its competitors to dryness and ultraviolet radiation that prevail in the foam.

One reason why there is an enormous increase of *Nocardia* during foaming is its ability to switch from a "K"-growth strategy to a "μ"-growth strategy. Most bacteria are K-growth strategists; that is, they can efficiently take up nutrients so that their growth rate is very steep when high concentrations of nutrients are available. As the level of nutrients diminishes, the growth curve levels off and then begins to decline. Those that are μ-growth strategists have a gradual growth curve but can reach much higher growth rates over a period of time (Figure 7-32). They are also adaptable to local and seasonal fluctuations of nutrients and can form resting stages during starving periods. The ability to switch from one strategy to another allows *Nocardia* to maintain a stable background population when nutrient conditions are restricted and competition by other organisms is strong.

Nocardia is generally dormant in colder temperatures but grows quickly in warmer temperatures. Once this filament gains dominance in the treatment system, it forms a stable foam that is difficult to control.

What makes this filament so difficult? In its early growth stage, *Nocardia* is harmless in the treatment system and does not

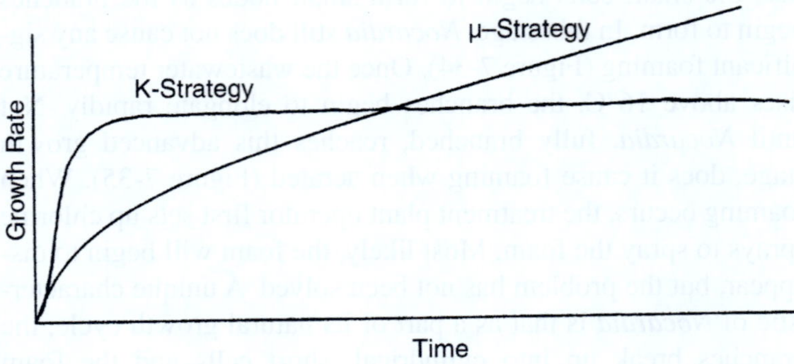

Figure 7-32 Growth curve. *Nocardia* can switch between strategies.

cause foaming. In this stage, it is unrecognizable as *Nocardia*. Instead of the classic branched filament, it lives as small, cylindrical, short, gram-positive (purple) cells (Figure 7-33) and can go undetected. As the temperature of the wastewater begins to

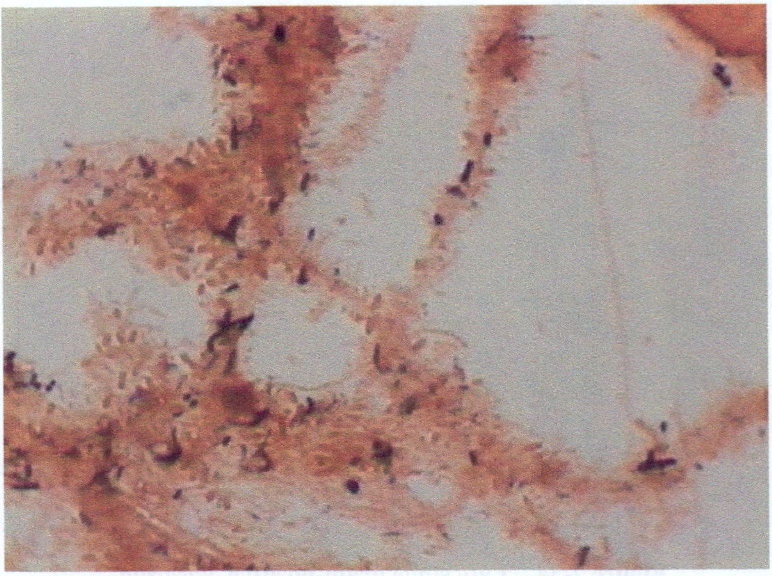

Figure 7-33 Gram-positive early stage *Nocardia* (Gram stain 1000× brightfield).

rise, the small cells begin to form small nodes as the branches begin to form. In this stage, *Nocardia* still does not cause any significant foaming (Figure 7-34). Once the wastewater temperature rises above 16°C, the branches begin to elongate rapidly. Not until *Nocardia*, fully branched, reaches this advanced growth stage, does it cause foaming when aerated (Figure 7-35). When foaming occurs, the treatment plant operator first sets up chlorine sprays to spray the foam. Most likely, the foam will begin to disappear, but the problem has not been solved. A unique characteristic of *Nocardia* is that as a part of its natural growth cycle, the branches break up into cylindrical, short cells and the foam begins to dissipate. The short cells begin to grow and branch once again and the growth cycle starts again.

Consider this: The foam dissipates and the new cell "babies" increase in number. They settle into the mixed liquor and begin to form nodes again, which eventually turn into branches. The increased number of branches will create a greater amount of foam. An operator who thinks spraying chlorine was successful in controlling the foam may increase the chlorine dosage. However,

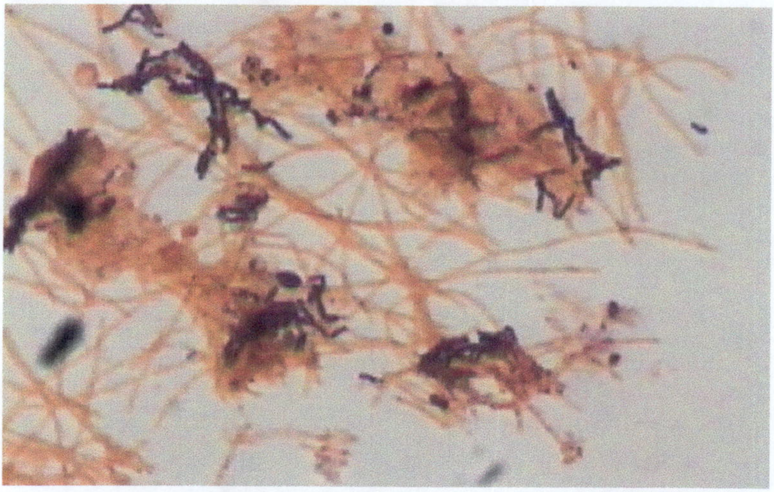

Figure 7-34 Late stage gram-positive *Nocardia*. At this stage branches begin to form. Insignificant foaming occurs at this stage (1000× brightfield).

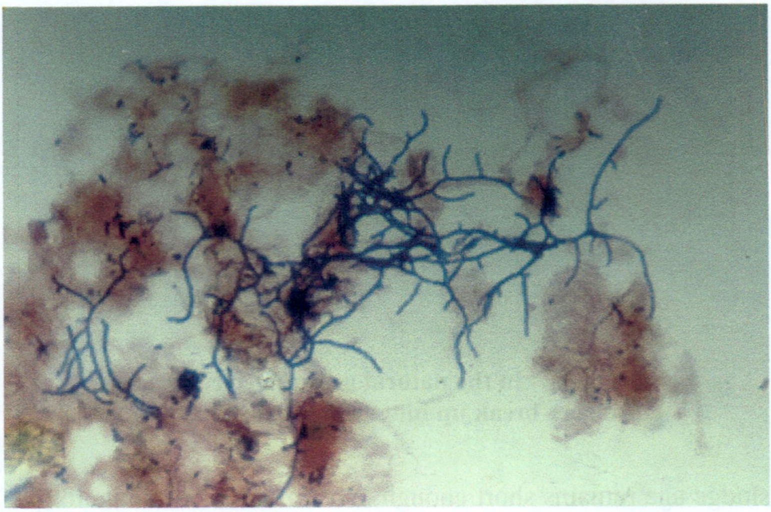

Figure 7-35 Gram-positive adult stage *Nocardia*. Fully branched, *Nocardia* causes foaming when aerated (1000× brightfield).

the growth cycles continue, and each generates more foam. So why is the foam out of control?

Chlorine works well for most filamentous bacteria because the filaments have much more surface area than the floc-forming bacteria or other protozoa. The chlorine can affect the filaments without adversely affecting the other microorganisms. But this is not so with *Nocardia*. Applying chlorine sprays to the foam will cause the *Nocardia* branches to prematurely break up into tiny cells and decrease the surface area. So in essence the operator is delivering "premature babies" that will eventually turn into branched filaments that cause more foam. So, what should the operator do?

Floc-forming bacteria grow rapidly in the activated sludge system, peaking in number early in the process. *Nocardia*, however, requires a much longer time to grow from its early growth stage to fully branched filaments. To a great degree, the sludge age determines whether *Nocardia* will remain in the early growth stage and cause no foam or mature to fully branched filaments and cause foaming in the system (Figure 7-36). As long as the

Figure 7-36 In the natural growth cycle, *Nocardia* branches break up into cylindrical short cells.

sludge age remains short enough, *Nocardia* will continue to live as harmless, small cells.

What about all the foam on the tanks, in the wet wells, and on the clarifiers? As long as the foam remains on the surface of the aeration basin, clarifiers, and wet wells, *Nocardia* will continue its growth cycle. The foam, once accumulated, also provides a source of seed for Nocardial growth. So, the foam must be sucked or skimmed off to prevent *Nocardia* from multiplying. Most importantly, the skimmed off foam should not be put back into the head of the treatment system but should be land applied or digested.

Eliminating foaming problems caused by *Nocardia* requires time, patience, and work. By decreasing the amount of time *Nocardia* remains in the aeration basin, you prevent it from maturing into branching filaments, and removing the foam from the system will prevent it from multiplying. After one or two summers, your system should be foam free!

Algae and Fungi

Algae are photosynthetic organisms (which contain chlorophyll) and obtain their energy from the sun and their carbon from carbon dioxide. Their size ranges from 1 micron to many meters. Algae are generally beneficial in lagoon treatment systems

because they supply oxygen for the bacteria in the treatment process. Algae generally do not cause problems in activated sludge treatment systems.

Fungi, though similar to algae, do not contain chlorophyll and require preformed organic matter as energy and carbon sources (e.g., sugars, fat, protein, and other carbohydrates). Fungi, ranging in size from a few microns to several centimeters, grow either independently by feeding on decaying matter or in association with plants and animals. They are rarely seen dominating in the activated sludge treatment system. Significant amounts of fungi in the treatment system usually indicate problems associated with pH changes and older sludge.

References

Eikelboom, D.H. 1977. Identification of Filamentous Bacteria in Activated Sludge. *Program Water Technology,* 8:153.

Jenkins, D., Richard, M., and Daigger, G. 1993. *Manual on the Causes and Control of Activated Sludge Bulking and Foaming.* Chelsea, Mich.: Lewis Publishers.

Richard, M. 1989. *Activated Sludge Microbiology.* Alexandria, Va.: The Water Pollution Control Federation.

Chapter 8

Microbiology and Process Control

Large wastewater treatment plants that handle hundreds of millions of gallons of water each day, in most cases, have enough staff to do only laboratory analyses. Unfortunately, this is not the case with smaller treatment facilities. Most wastewater treatment facilities employ one or two operators, whose duties involve treating wastewater and drinking water, plowing snow, and mowing the grass. To properly operate the wastewater treatment facility and meet regulatory requirements, certain laboratory analyses must be performed. Adding tests, especially ones that use a microscope, may be the straw that breaks the operator's back. Keeping this in mind and considering the value of microscopic evaluations in the operation of the treatment system, following are some simple microscopic observations the operator can make that will require minimal time and effort.

Dispersed Bacteria

Ideally, when observing a sample of mixed liquor under a microscope, one would see a variety of protozoa (mostly ciliates), along with well-defined compact floc surrounded by clear fluid. In a well-operated system, most of the dispersed bacteria have developed into floc and the ciliates have removed the remaining

dispersed bacteria and clarified the effluent. By observing the fluid surrounding the floc, the operator can determine whether the floc is forming properly. Ideally, the surrounding fluid should be clear of dispersed growth (Figure 8-1). A great deal of dispersed growth in the surrounding fluid with weakly formed floc may mean that something has interfered with proper floc formation (see chapter 4). If the floc is well formed, influent toxicity may have affected the ciliates' ability to clarify the effluent.

Toxicity

The higher life forms are the first to be affected whenever a toxic substance enters the treatment plant. If a fresh sample of mixed liquor contains dead rotifers and inactive stalked ciliates, something toxic has probably entered the plant. Loss of BOD removal and an unusually low oxygen use also indicate toxicity (dead microorganisms do not remove BOD).

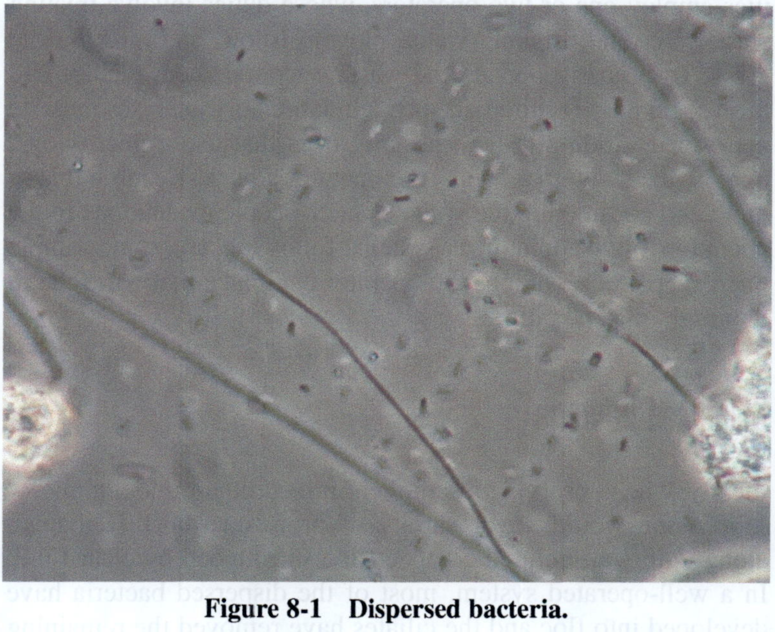

Figure 8-1 Dispersed bacteria.

Slime Bulking

As discussed earlier, slime bulking can result when there are insufficient nutrients for the bacteria. Under nutrient-deficient conditions, bacteria secrete excess exocellular lipopolysaccharide. This slime has a buoyant effect on the floc particles and causes poor settling. Bacteria will also produce excess slime in the presence of significant amounts of organic acids. These acids, usually produced under anaerobic conditions, can be recycled into the plant when anaerobic digester supernatant is slug-dosed into the recycle stream. If this is the case, a supernatant should be introduced into the system in smaller doses over a longer period of time. In some cases, when sludge is held too long in the primary clarifier, organic acids produced in the sludge can enter the aeration basin through the primary effluent.

If a significant amount of this slime is present in the mixed liquor, it can be detected without a microscope. The sludge will have a slimy consistency, as if it had been mixed with polymer. Otherwise, a simple procedure called the India Ink Test can be used to detect the excess slime.

India Ink Test

1. Place a small drop of mixed liquor on a clean, grease-free slide.
2. Add a small drop of water-soluble India ink. (This can be purchased at an art supply store.) Cover with a cover slip. Be sure to remove excess fluid from around the slide by touching a piece of paper towel to the edge of the slide to draw off the excess.
3. Place a drop of immersion oil directly on top of the cover slip and view the slide with the 100× oil immersion lens.
4. Observe the sample with brightfield and the maximum light possible. Do not use the phase contrast condenser.

When India ink is added to a drop of mixed liquor, the carbon black particles penetrate the floc from outside to inside. The slime around the bacterial cells prevents the particles from penetrating the floc. The carbon black particles cannot penetrate through this

slimy, polymer-like substance. If the slime is not present, the floc will appear black because the ink has penetrated it. If the slime is present, the floc will appear white (Figure 8-2).

Protozoa Count

As environmental components of the activated sludge change, one type of microorganism is replaced by another. The microorganism best suited for the environment will emerge until the environment changes. Changes in pH, DO, temperature, nutrients, competition, etc., all determine which species will dominate. The protozoan species that are most dominant in the treatment system indicate which conditions are most dominant. Protozoan species dominance should not be relied on solely to troubleshoot wastewater treatment conditions; however, this information is helpful in assessing the conditions of the activated sludge process.

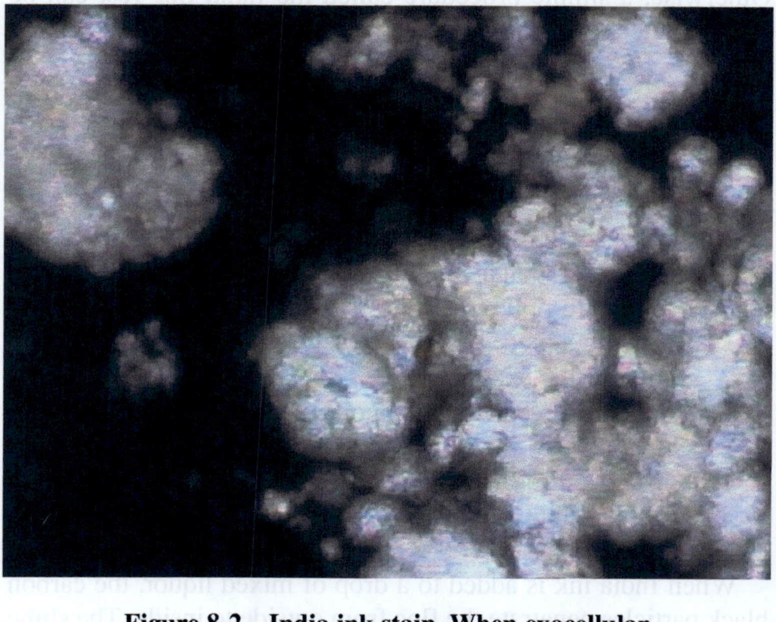

Figure 8-2 India ink stain. When exocellular lipopolysaccharide is present in excess, the ink cannot penetrate the floc.

This protozoa count procedure is not designed to determine the total number of each type of protozoan in the system. Instead, the relative numbers of one type in comparison to another type are important. In other words, the purpose is to determine which species seem to dominate. The count will examine protozoa in the following categories:

- Amoebae
- Flagellates
- Free-swimming ciliates
- Crawling ciliates
- Stalked ciliates
- Metazoa (rotifers, nematodes, waterbear, bristle worms, etc.)

In a well-operating system, the ciliates will most likely be the dominant species. Ciliates dominate when most of the nutrients have been removed from the wastewater. In a poorly operated system, amoebae and flagellates will probably dominate. Amoebae and flagellates can compete for dominance only when plenty of nutrients remain. Stalked ciliates and metazoa will dominate in older systems, primarily because they can compete when few nutrients are left and they can feed on other protozoa.

The Procedure

Collect a fresh, well-mixed, representative sample of mixed liquor from the discharge end of the aeration basin. If the concentration of the mixed liquor is quite high, dilute the sample. Always use the same dilution if you plan to do routine counts. Ideally, you should observe the sample under the 20× objective with the phase contrast condenser. Using a smaller objective (10× or lower) will make it difficult to see the smaller flagellates, and using a larger objective (40× or higher) will narrow your field of view.

1. Place one drop of mixed liquor on a clean, grease-free slide and cover with a cover slip.

2. Scan the slide with five passes. Focus the objective at the top left corner of the cover slip. Move down the cover slip count and record the number of each type of protozoa that you see. This completes one pass. Move over to the right slightly and move up the cover slip, keeping a running total of the number of each type of protozoa. This will complete the second pass. Move up and down the slide until you have completed five passes (Figure 8-3).
3. Record the running total number of each type of protozoa (see appendix A for a sample protozoan count worksheet). If you encounter colonies of stalked ciliates, you must count (or estimate) the total number of heads. The more heads, the older the sludge. For best results, scan three or more slides and average the number of each protozoa type.
4. Calculate relative dominance:
   a. Count and record the total number of protozoa in each category. Record the counts from all three slides (see example in Table 8-1).
   b. Calculate the average number of each organism from all three slides.
   c. Calculate the percentage by dividing the average number of each organism by the total number of microorganisms and multiply by 100.

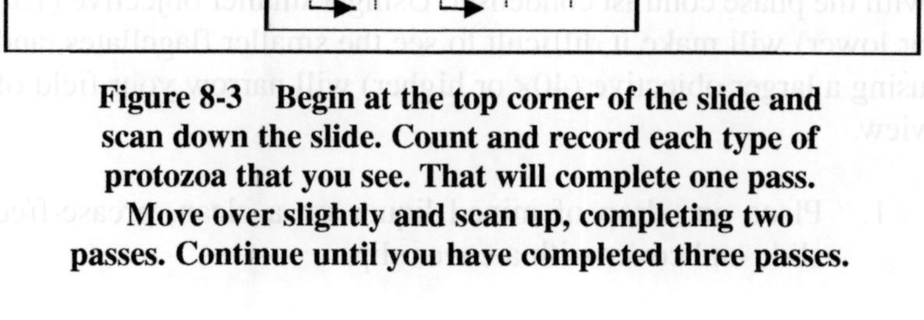

Figure 8-3 Begin at the top corner of the slide and scan down the slide. Count and record each type of protozoa that you see. That will complete one pass. Move over slightly and scan up, completing two passes. Continue until you have completed three passes.

Table 8-1 Categories of Protozoa

Organism	*Slide #1*	*Slide #2*	*Slide #3*	*Average*		*Percent*
Amoebae	6	3	1	3	4	14.5
Flagellates	13	6	6	8	10.5	
Free-swimming ciliates	15	12	7	11	14.5	81.6
Crawling ciliates	22	15	20	19	25	
Stalked ciliates	40	32	23	32	42	
Metazoa (rotifers, nematodes, etc.)	5	1	2	3	4	3.9
Total	101	69	59	76		

d. Group amoebae and flagellates together and combine the percentage. Group all the ciliates together, combine the percentage, and calculate the percentage of metazoa.

Amoebae and flagellates are grouped together because they indicate young sludge or incomplete treatment. In a well-operated system, ciliates should dominate. Higher numbers of metazoa are associated with older sludge.

Protozoa counts are not exact and should not be relied on solely to determine treatment system conditions. They are helpful, however, in assessing conditions in the aeration basin.

In the world of wastewater treatment process control, there are no absolutes. Wastewater is not homogeneous. There is no way to predict the amount of each substance that enters the treatment system. The system is affected by temperature, the availability of oxygen, pH, microbial competition, and other influences. You cannot say that an F/M ratio of 0.3 that is good for one treatment plant will work well in another. Therefore, you must know what works well for your treatment system. Keep a process chart and record information on plant processes when the system is running well (see appendix B for a sample process control chart). You can modify the chart to suit the processes in your system. For example, record the following parameters daily or weekly:

- Influent BOD/COD—This should represent the influent load to the aeration basin, not to the primary clarifier.
- Aeration basin DO—Measure this daily at the influent end of the basin and again at the discharge end. The change in DO indicates how well BOD is being removed. Generally, the DO at the influent end should be lower than the DO at the discharge end. The influent end usually contains the highest concentration of BOD and microorganisms that consume more oxygen. At the discharge end, the oxygen consumption should be less. If the aerators are evenly distributed and the DO at the influent end of the basin is the same as the DO at the discharge end of the basin, the BOD is probably not being removed properly.
- MLSS and MLVSS—Measure MLSS concentration daily. MLVSS can be measured once per week and is generally used to estimate the concentration of microorganisms and determine the F/M ratio. You can also use a spin centrifuge to calculate the mass of solids in the aeration basin.
- F/M ratio—Determine this ratio by using the influent BOD or COD to the aeration basin as food and the MLVSS concentration as the concentration of microorganisms.
- 30-minute settled volume—A test that shows how well the sludge is settling.
- Protozoa count—To determine which protozoan group is dominant, you can do the protozoa count described in this chapter. If you are interested in looking at trends, you must factor in the MLVSS. A higher concentration of mixed liquor will probably give a higher protozoa count. To determine the protozoa count per milligram of mixed liquor, divide the count by the MLVSS and record this number in the process chart instead.

Once you are familiar with the process control parameters that are best suited for your treatment plant, you will have a baseline on which to base any changes to your system performance. As you begin to notice changes in your plant performance, you will be able to review the information to see where any significant changes occurred. Changes in one parameter can affect others. The important thing is to understand your own treatment plant.

Microorganisms are natural assets to the health of any community. They are necessary for our physical health; in the synthesis of products such as acetone, glycerin, organic acids, enzymes, and alcohols; and in the manufacture of many drugs. Even our food industry relies on microorganisms to help produce vinegar, sauerkraut, pickles, alcoholic beverages, buttermilk, cheeses, yogurt, and bread. Microorganisms are essential to waste disposal.

Your role as wastewater operator is also essential to the health of your community. Your underrated role is more than a man or woman with a plunger. Wastewater treatment is a sophisticated, scientific process that requires technical knowledge and skill. As this book applauds the microorganisms that are responsible for keeping us from drinking and swimming in our own wastes, it applauds the operators who tend to their needs and who know that if you treat the microorganisms right, they will treat the water right!

Appendixes

APPENDIX A
PROTOZOAN COUNT WORKSHEET

ORGANISM	Slide #1	Slide #2	Slide #3	Average	Percent	
Amoeba						
Flagellate						
Free-Swimming Ciliates						
Crawling Ciliates						
Stalked Ciliates						
Metazoa (Rotifers, Nematodes, etc.)						
			Average Total			

APPENDIX B
PROCESS CONTROL CHART

			Basin DO					% Protozoa					
Date	Influent BOD/COD	30-Minute Settling	**Inf**	**Eff**	MLSS	MLVSS	F/M Ratio	Amoeba	Flagellates	Free-Swimming Ciliates	Crawling Ciliates	Stalked Ciliates	Metazoa

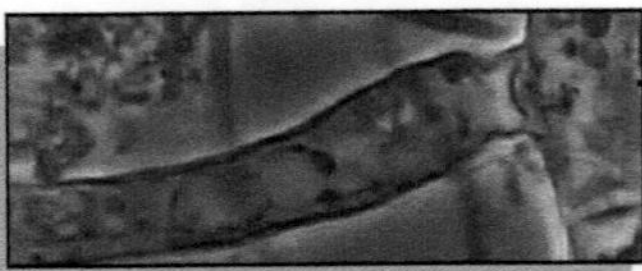

Appendix C

The images that appear in this appendix correspond to those on the CD-ROM, which is included with this book.

BACTERIA

FILAMENTOUS BACTERIA

FILAMENTOUS BACTERIA

(continued)

METAZOA

PROTOZOA

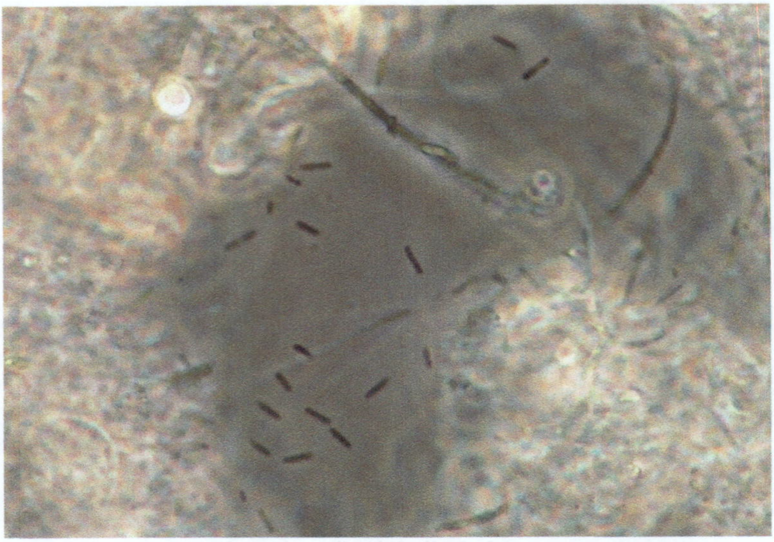

1. Bacillus-shaped bacteria
Bacillus-shaped bacteria are shaped like round-ended rods, squares, or rectangles (Wet mount, 1000× phase contrast).

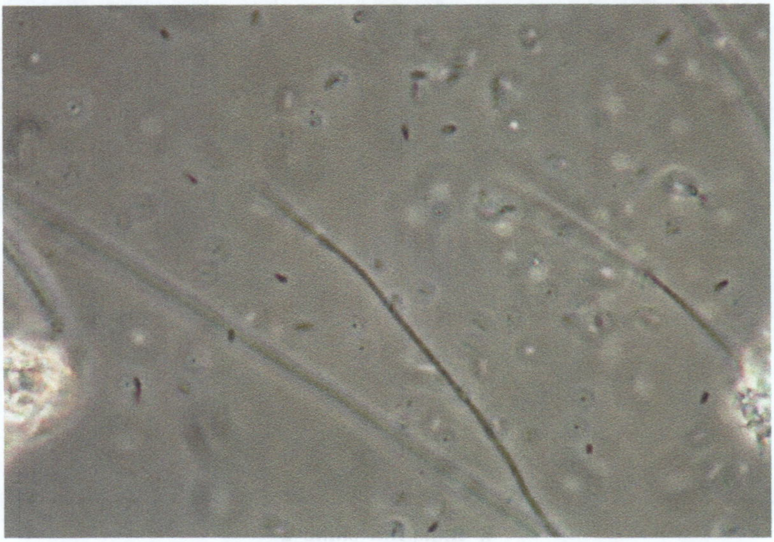

2. Dispersed bacteria
Free-swimming, actively growing bacteria (Wet mount, 1000× phase contrast).

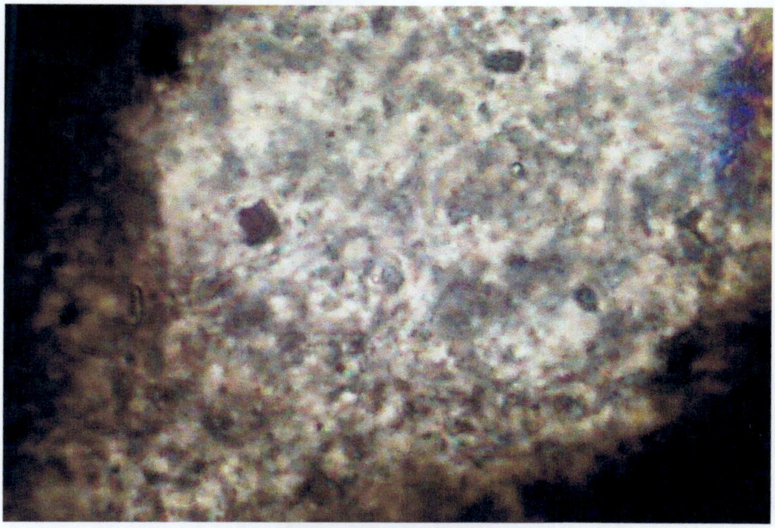

3. India Ink_slime_1
A positive India ink test indicating the presence of excess exocellular lipopolysaccharide (slime). The white area of the floc indicates slime is present (India ink stain, 1000× brightfield).

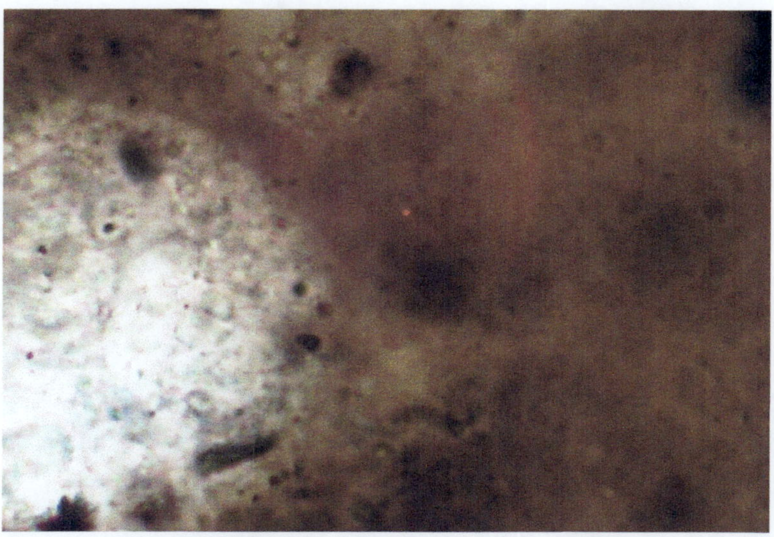

4. India Ink_slime_2
A positive India ink test indicating the presence of excess slime in the floc as well as in the bulk fluid (India ink stain, 1000× brightfield).

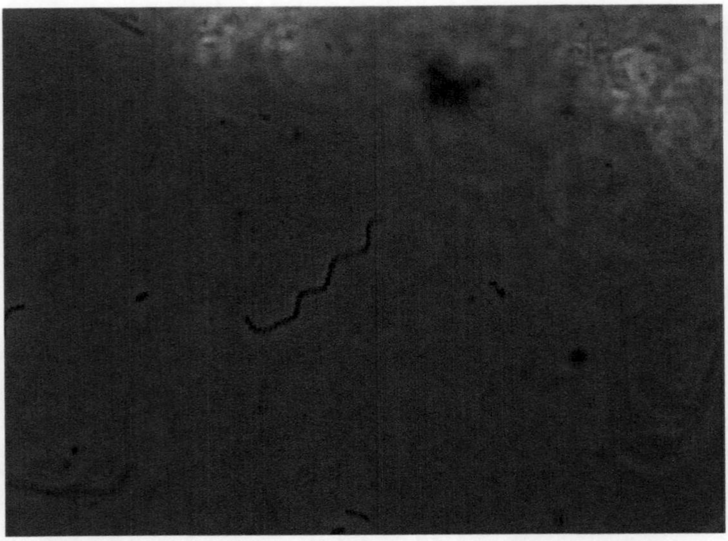

5. Spirochete_1
Spirochete, a double corkscrew shaped bacterium with a very flexible body (Wet mount, 1000× phase contrast).

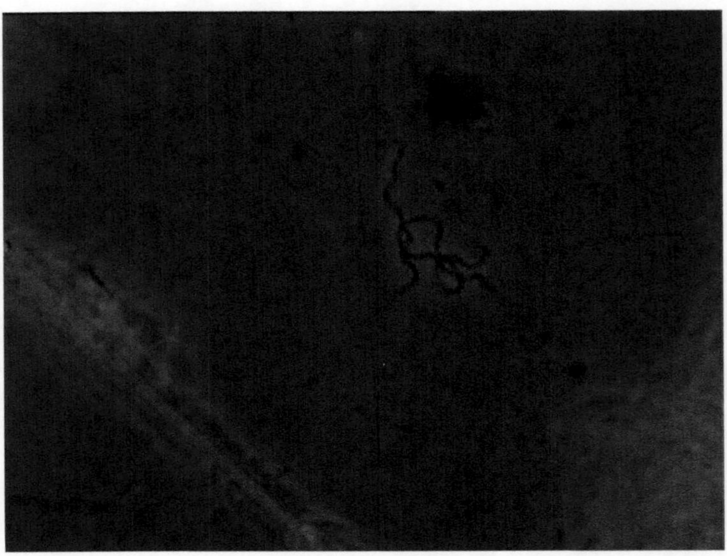

6. Spirochete_2
(Wet mount, 1000× phase contrast).

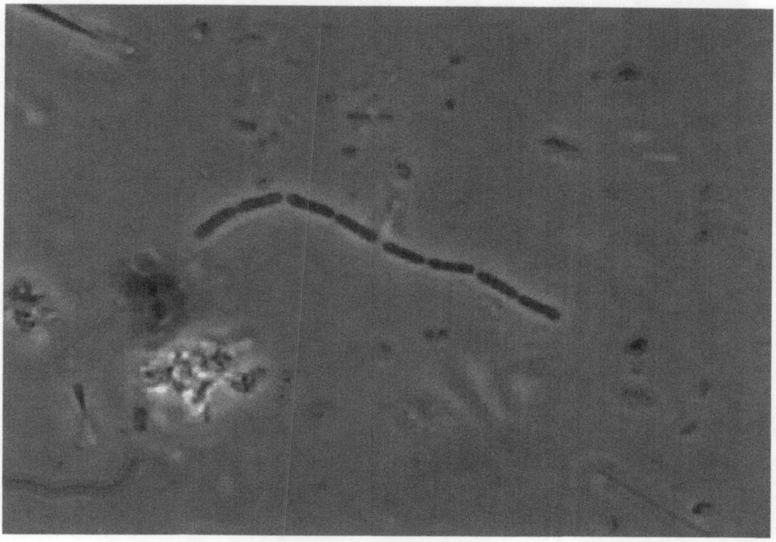

1. Bacillus spp_1
Large rod-shaped bacteria that form a chain and stain gram-positive (Wet mount, 1000× phase contrast).

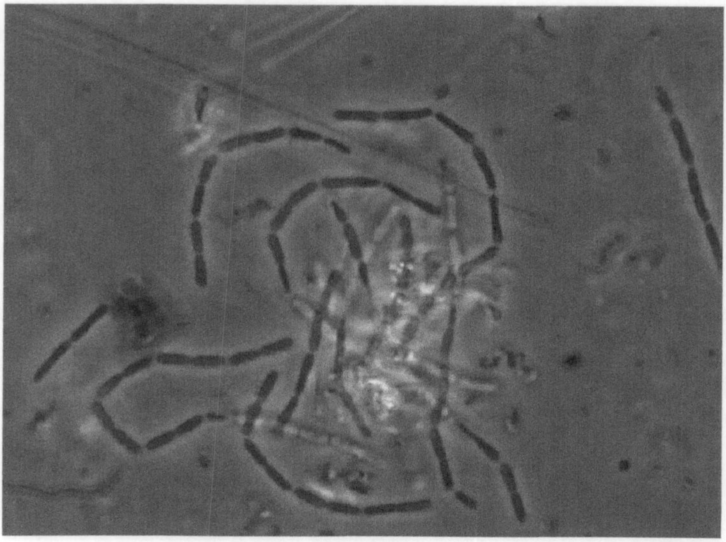

2. Bacillus spp_2
(Wet mount, 1000× phase contrast).

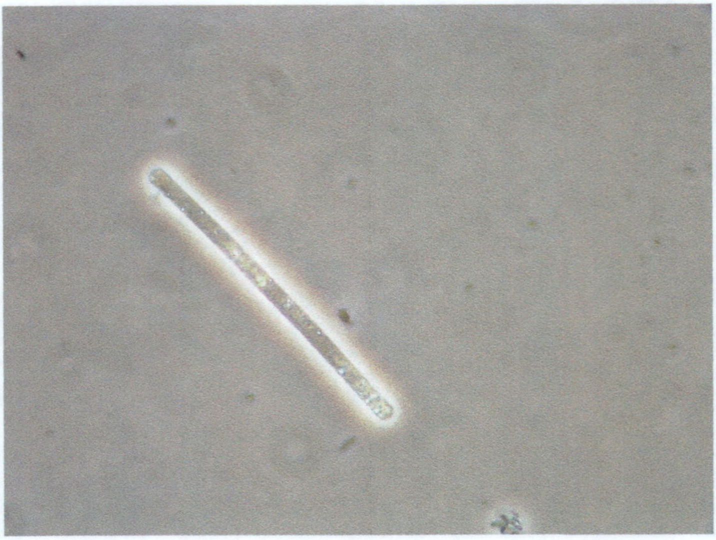

3. Beggiatoa_1
Beggiatoa filament containing sulfur granules (Wet mount, 1000× phase contrast).

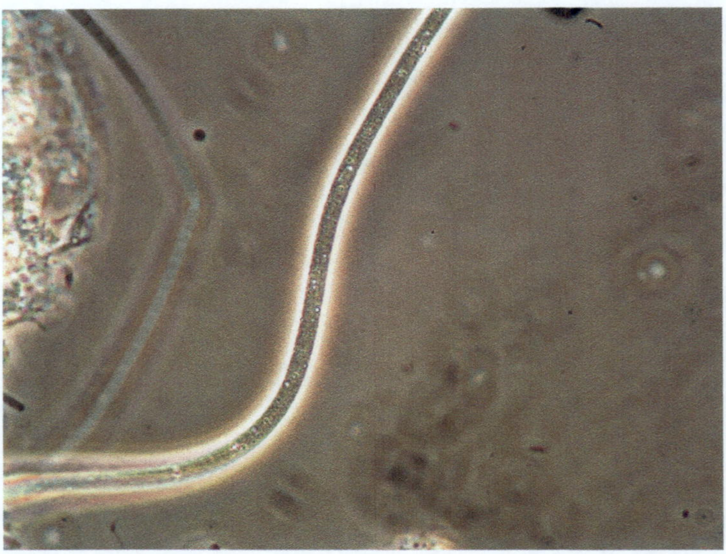

4. Beggiatoa_2
Beggiatoa filament containing sulfur granules. Beggiatoa can occur as a single cell or in a filament ranging from 1 to 50 microns long (Wet mount, 1000× phase contrast).

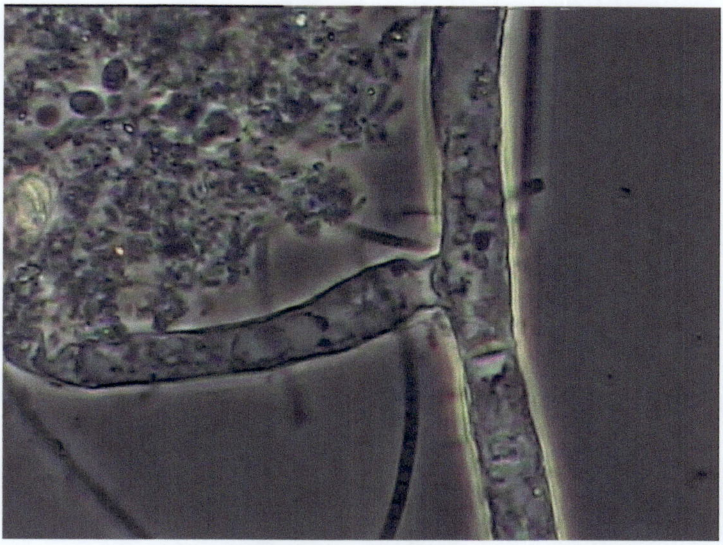

5. Fungus_true branching
Fungus exhibiting true branching
(Wet mount, 1000× phase contrast).

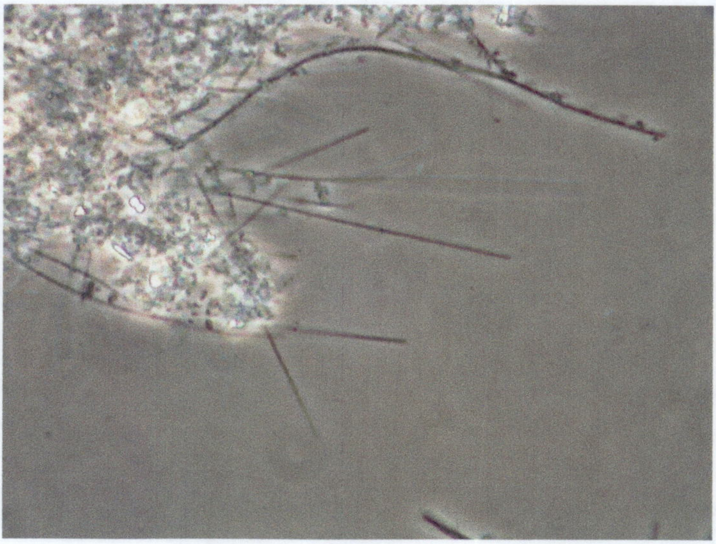

6. H. hydrossis_1
Haliscomenobacter hydrossis. A very thin, sheathed, straight filament that resembles pins in a cushion
(Wet mount, 1000× phase contrast).

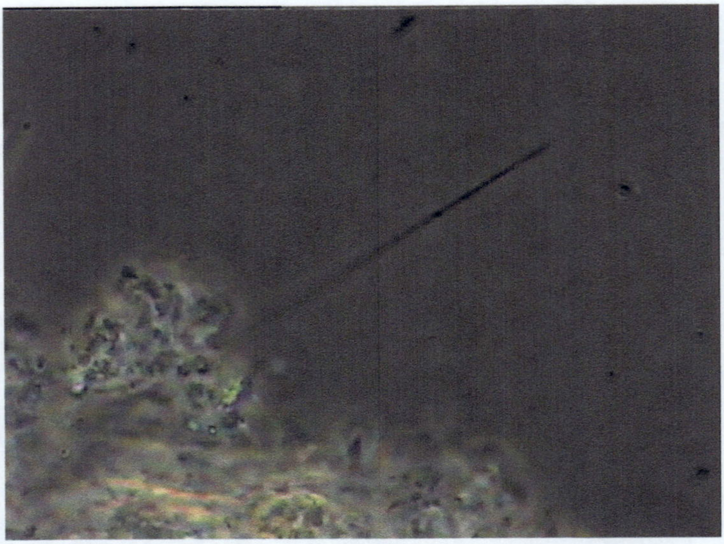

7. H. hydrossis_2
Haliscomenobacter hydrossis. The presence of a sheath on this filament is difficult to detect but is indicated by the missing space along the filament (Wet mount, 1000× phase contrast).

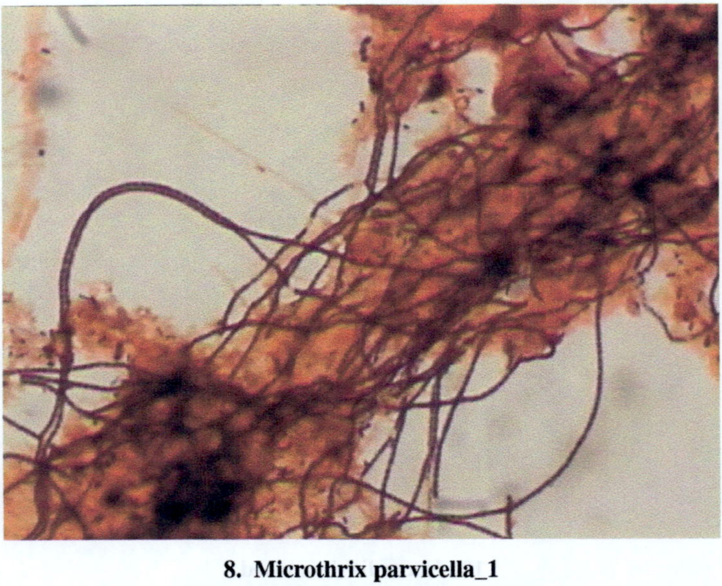

8. Microthrix parvicella_1
Gram-positive Microthrix parvicella (Gram stain, 1000× brightfield).

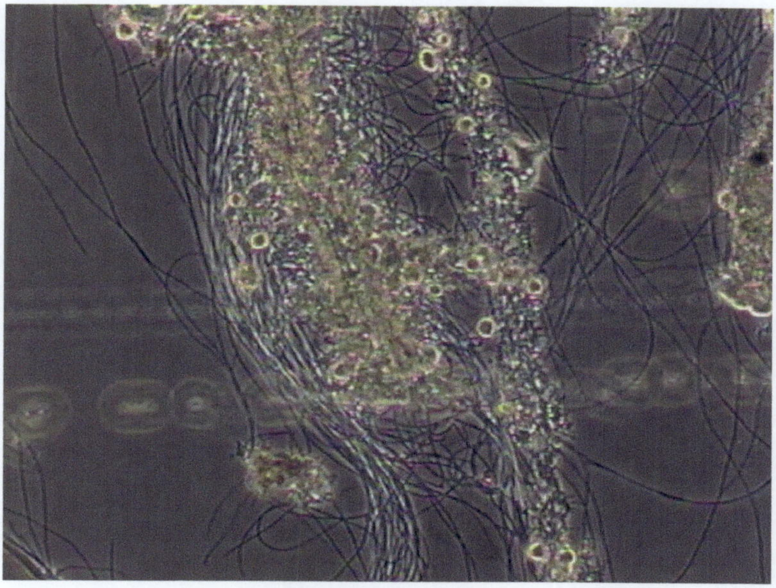

9. Microthrix parvicella_2
A wet mount of a foam sample containing excess amounts of Microthrix parvicella (Wet mount, 1000× phase contrast).

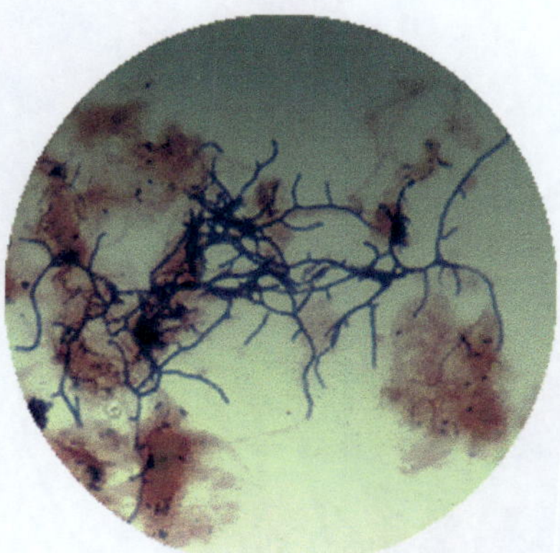

10. Nocardia_branched
Fully branched, adult-stage gram-positive Nocardia spp. (Gram stain, 1000× brightfield).

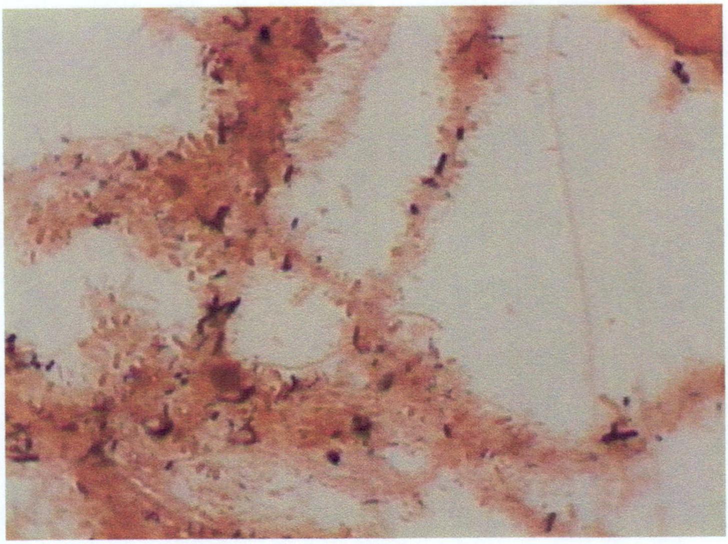

11. Nocardia_early stage
Gram-positive, early-stage Nocardia spp.
(Gram stain, 1000× brightfield).

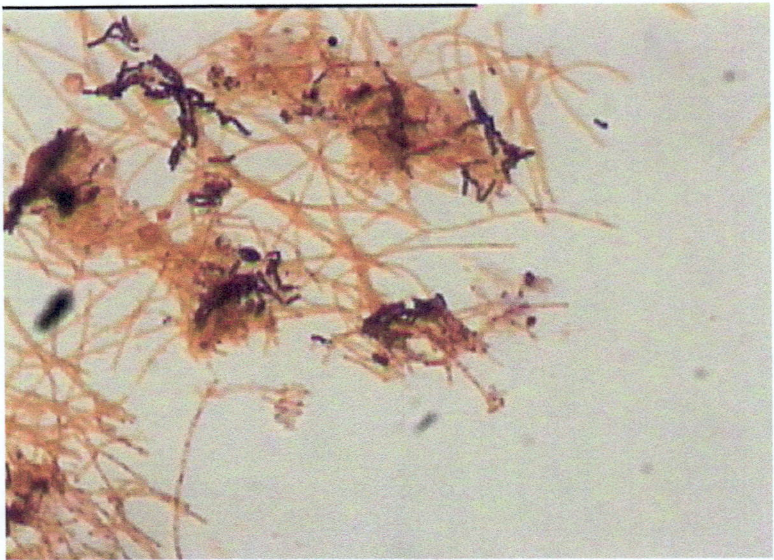

12. Nocardia_mid-stage
Gram-positive, mid-stage Nocardia spp.
(Gram stain, 1000× brightfield).

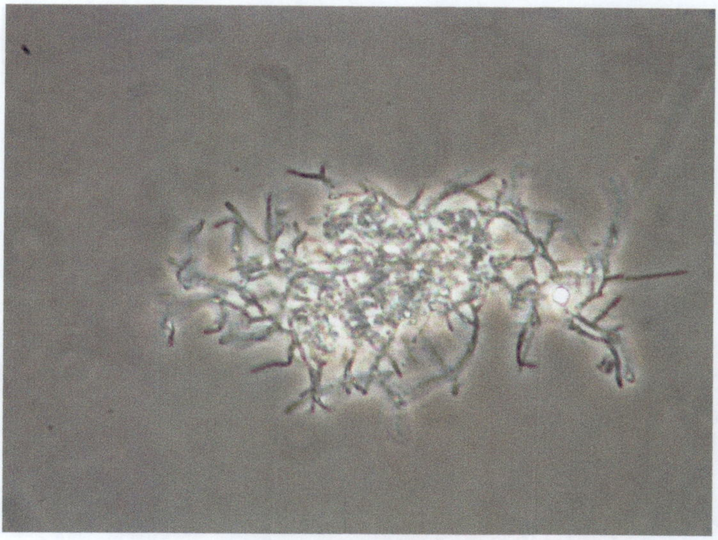

13. Nocardia_wet mount
Fully branched, adult-stage Nocardia
(Wet mount, 1000× phase contrast).

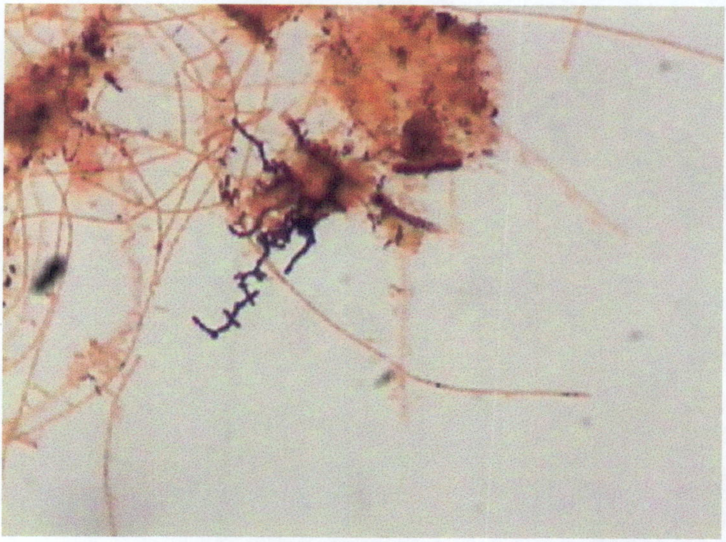

14. Nocardia_young adult
Nocardia spp. beginning to form nodes for branches
(Gram stain, 1000× brightfield).

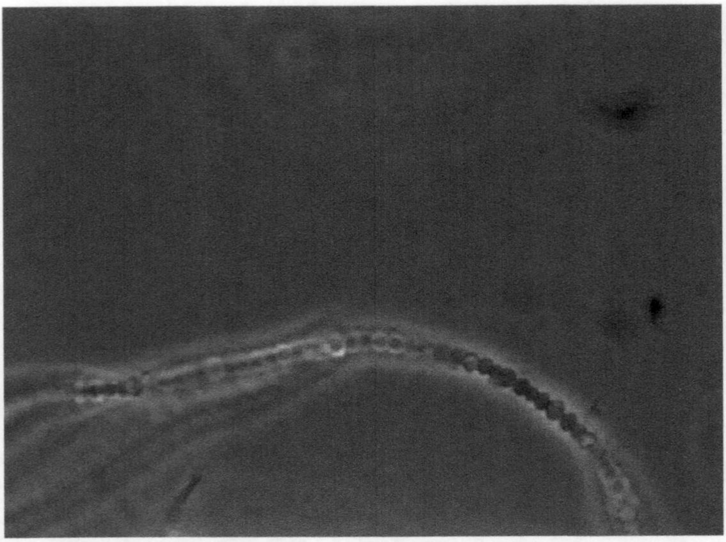

15. Nostocoida limicola II_1
(Wet mount, 1000× phase contrast)

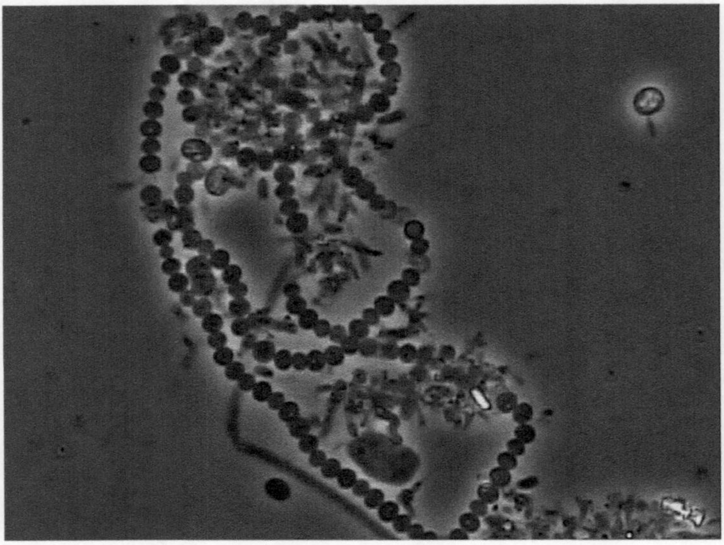

16. Nostocoida limicola II_2
(Wet mount, 1000× phase contrast)

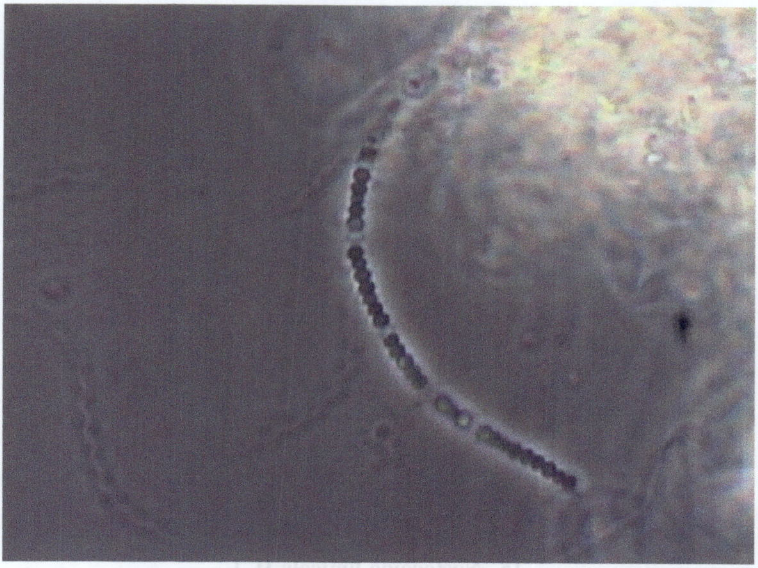

17. Nostocoida limicola II_3
(Wet mount, 1000× phase contrast)

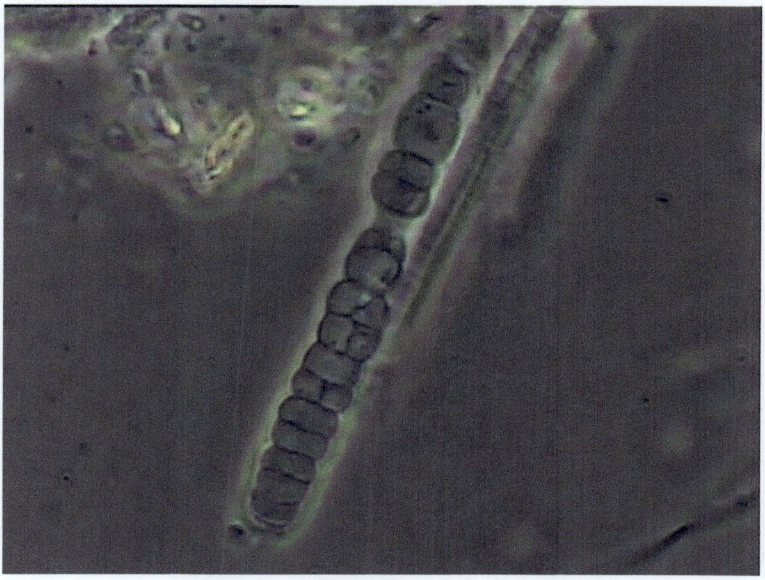

18. Nostocoida limicola III
(Wet mount, 1000× phase contrast)

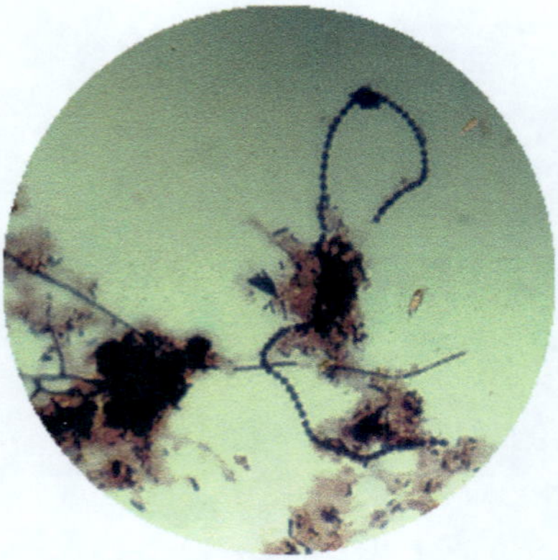

19. Nostocoida limicola-Gram
Gram-positive Nostocoida limicola
(Gram stain, 1000× brightfield).

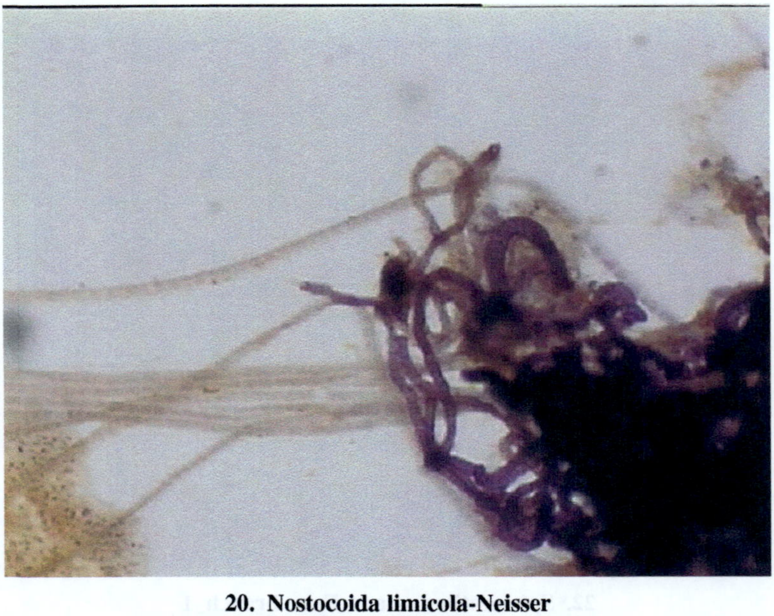

20. Nostocoida limicola-Neisser
Neisser-positive Nostocoida limicola
(Neisser stain, 1000× brightfield).

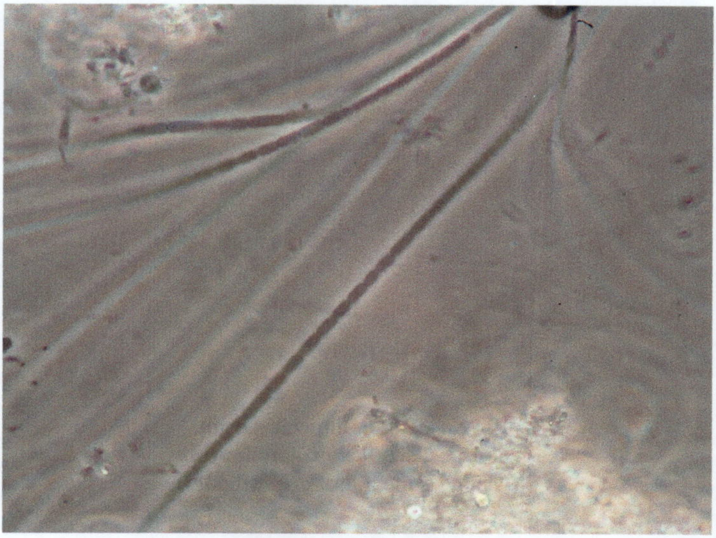

21. Sphaerotilus natans
Sphaerotilus natans. Sausage-shaped cells
(Wet mount, 1000× phase contrast).

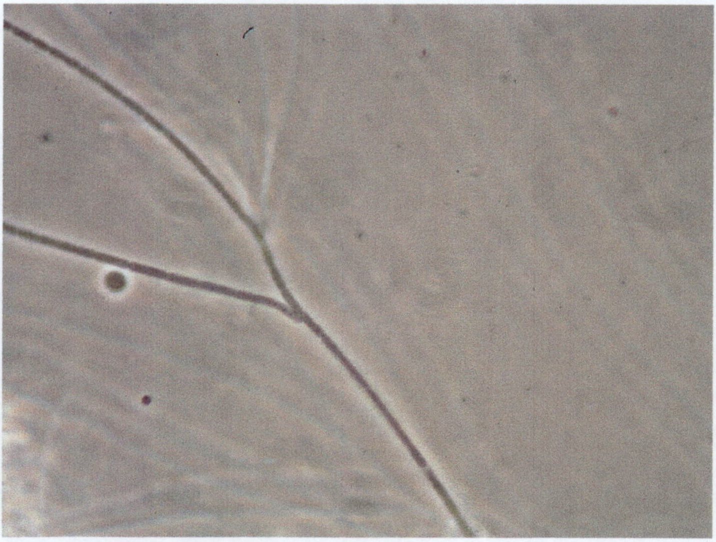

22. Sphaerotilus natans_False branch_1
Sphaerotilus natans. False branching
(Wet mount, 1000× phase contrast).

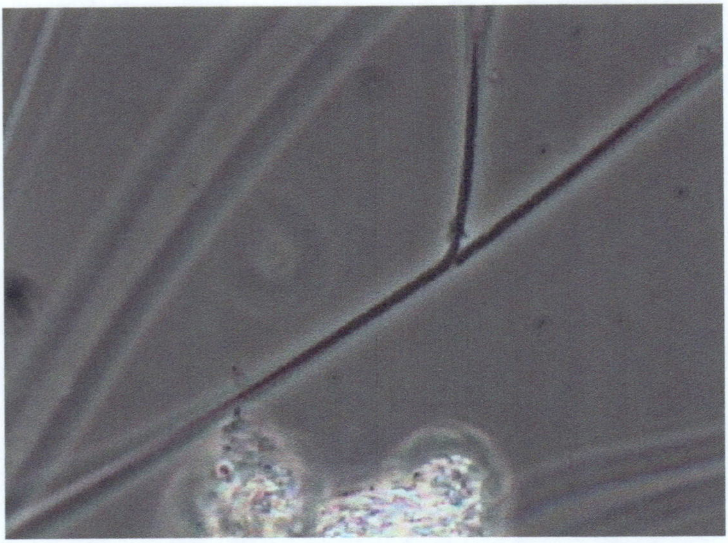

23. Sphaerotilus natans_False branch_2
Sphaerotilus natans. False branching
(Wet mount, 1000× phase contrast).

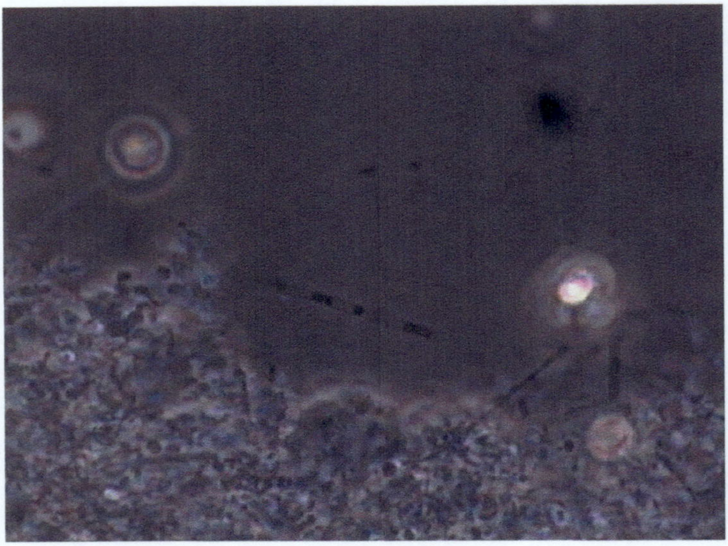

24. Sheath_1
Filament sheath (Wet mount, 1000× phase contrast).

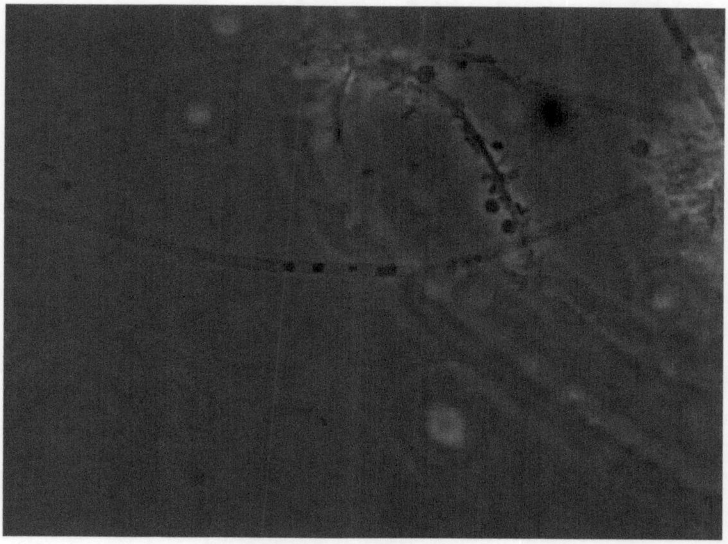

25. Sheath_2
Filament sheath (Wet mount, 1000× phase contrast).

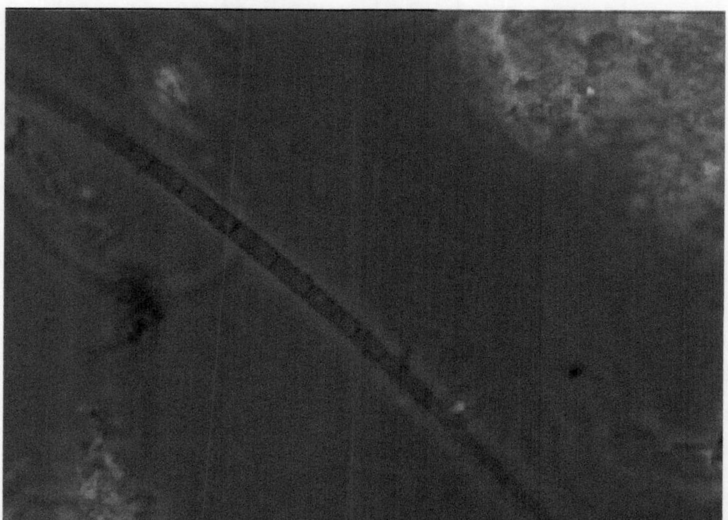

26. Thiothrix I
Thiothrix sp. Type 1 (Wet mount, 1000× phase contrast).

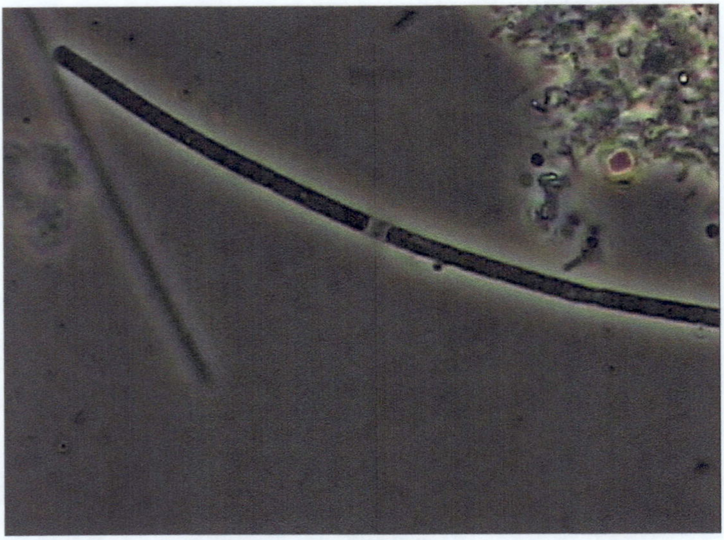

27. Thiothrix I_Sheath
Thiothrix sp. Type I. The sheath can be detected when there are missing cells along the filament (Wet mount, 1000× phase contrast).

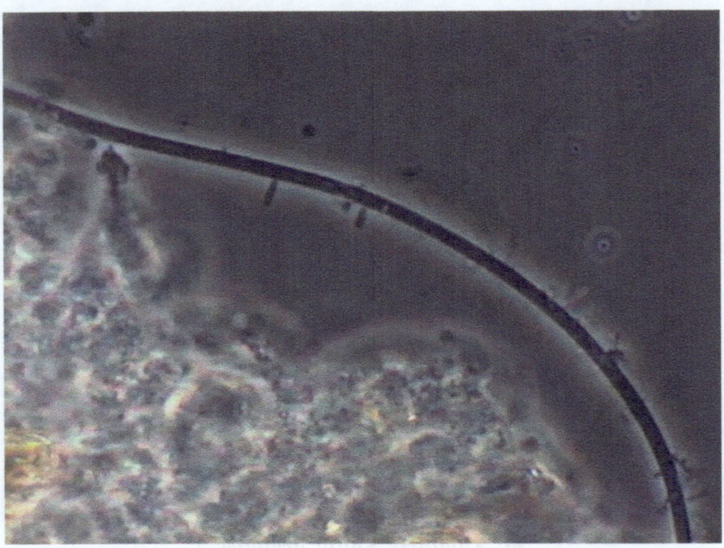

28. Thiothrix II
Thiothrix sp. Type II (Wet mount, 1000× phase contrast).

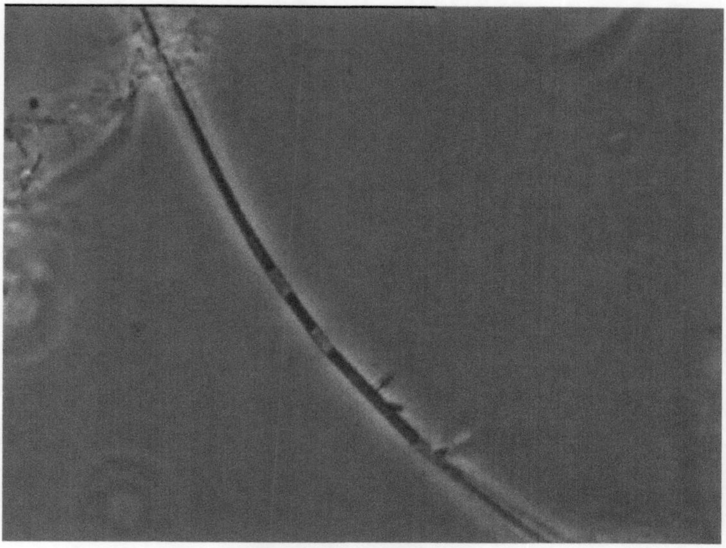

29. Thiothrix II_Sheath
Thiothrix sp. Type II. The sheath can be detected when there are missing cells along the filament (Wet mount, 1000× phase contrast).

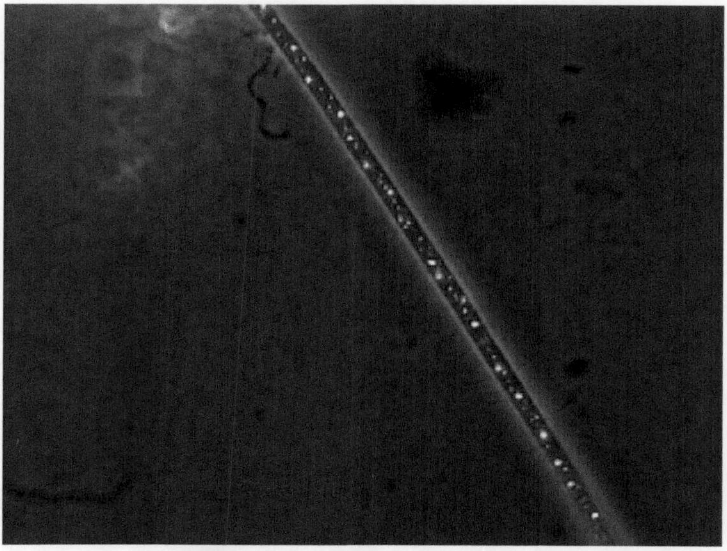

30. Thiothrix_Sulfur granules_1
Thiothrix sp. Sulfur granules are stored within the cells of the filament (Wet mount, 1000× phase contrast).

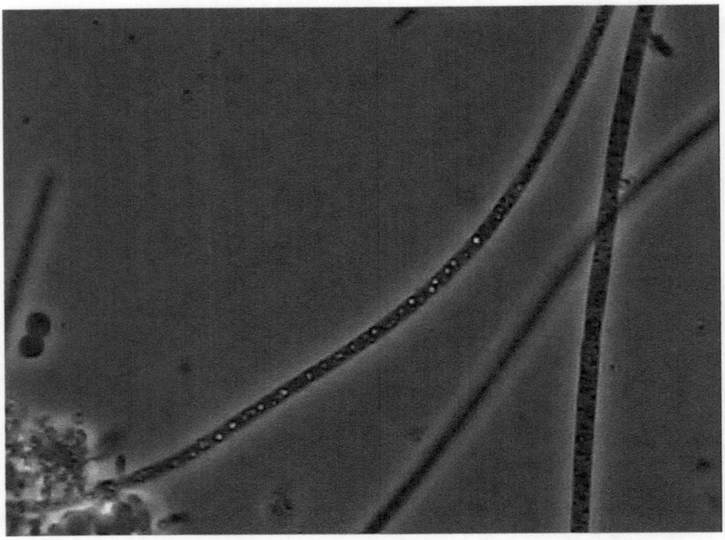

31. Thiothrix_Sulfur granules_2
Thiothrix sp. Sulfur granules are stored within the cells of the filament (Wet mount, 1000× phase contrast).

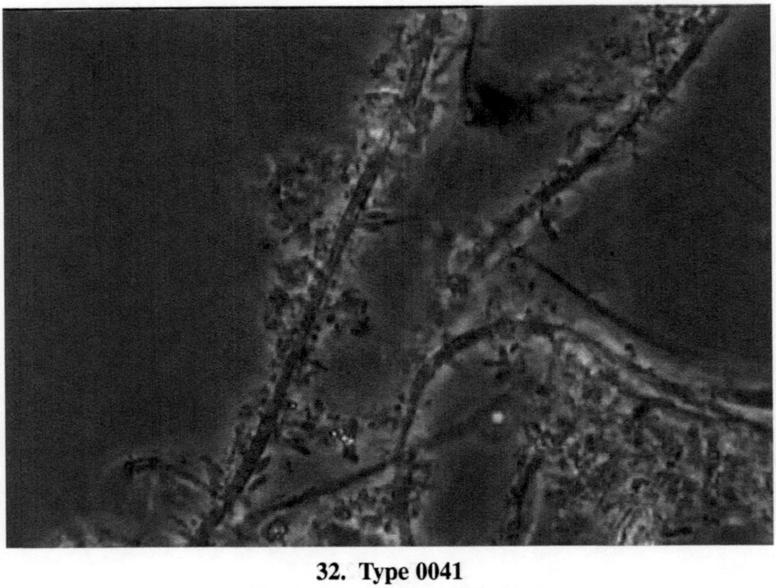

32. Type 0041
Type 0041 with heavy attached growth (Wet mount, 1000× phase contrast).

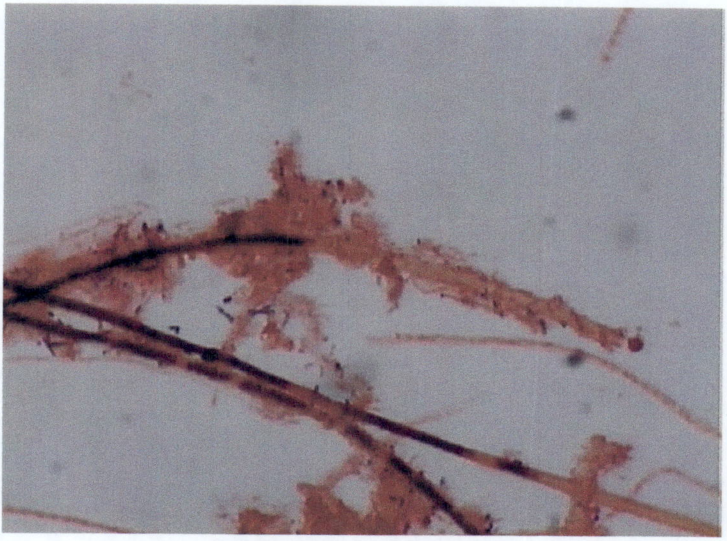

33. Type 0041_GramVariable
Type 0041 Gram variable (Gram stain, 1000× brightfield).

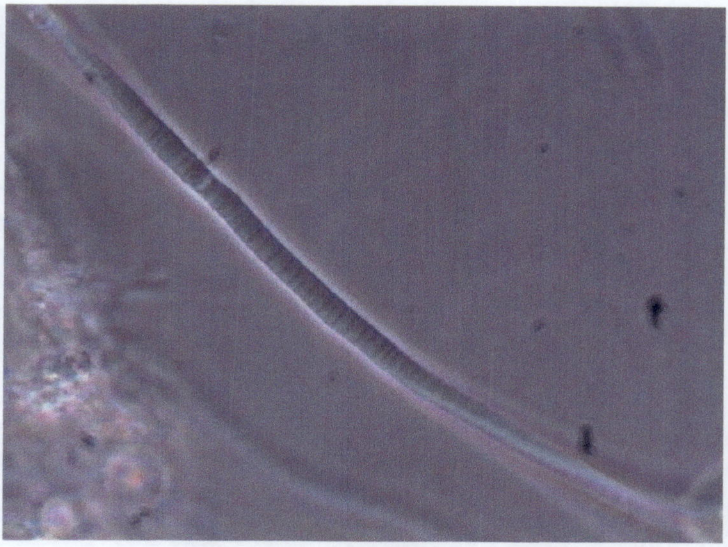

34. Type 021N
Type 021N discoid-shaped cells
(Wet mount, 1000× phase contrast).

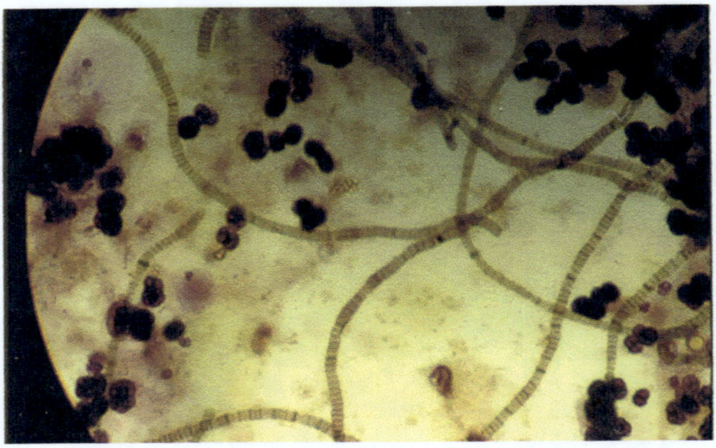

35. Type 021N_Neisser
Type 021N displaying a Neisser-negative reaction (Neisser stain, 1000× phase contrast).

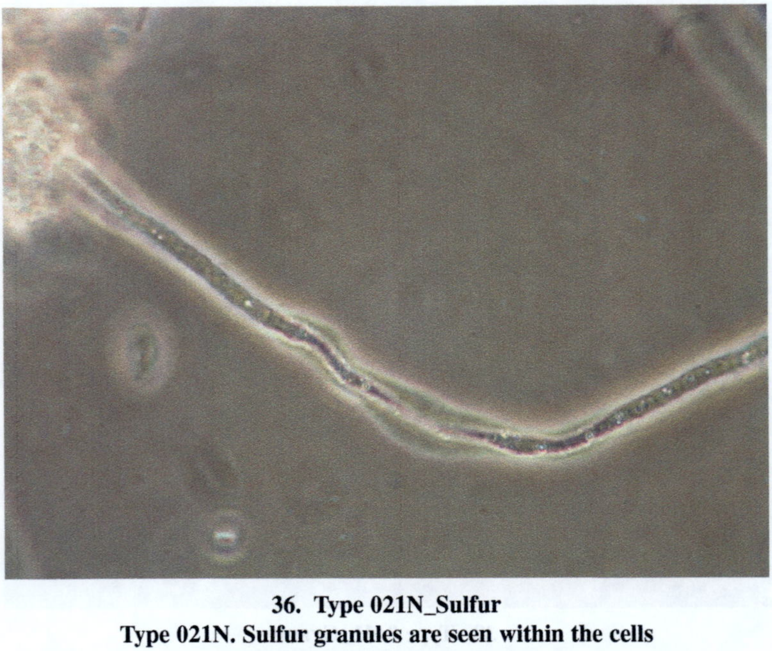

36. Type 021N_Sulfur
Type 021N. Sulfur granules are seen within the cells (Wet mount, 1000× phase contrast).

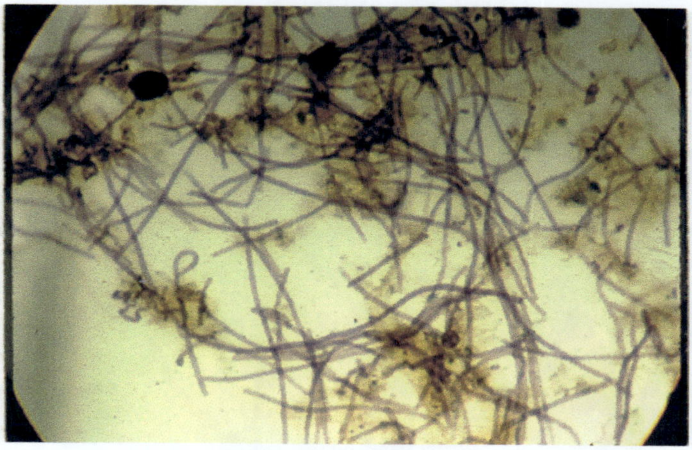

37. Type 0092_Neisser
Neisser-positive type 0092 (Neisser stain, 1000× brightfield).

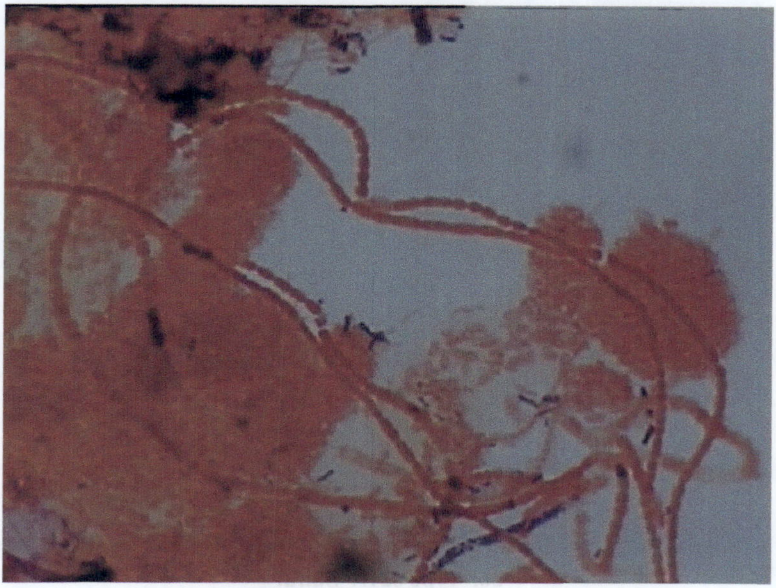

38. Type 0411_Gram
Gram-negative type 0411 (Gram stain, 1000× brightfield).

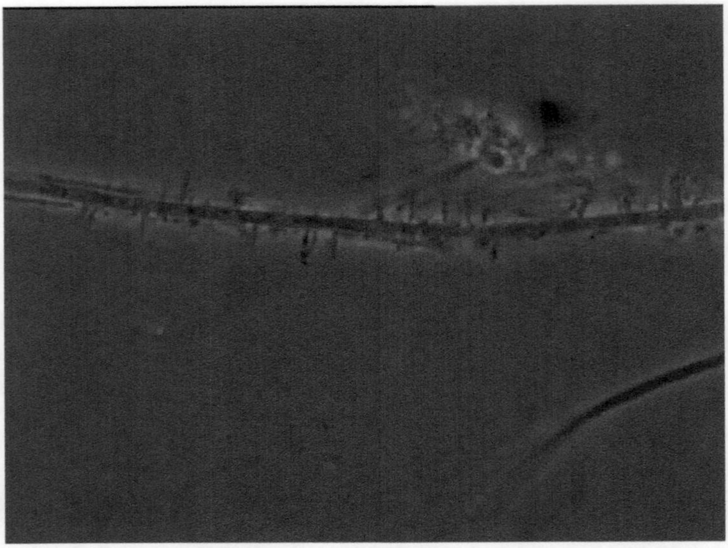

39. Type 0675_1
Type 0675 with heavy attached growth
(Wet mount, 1000× phase contrast).

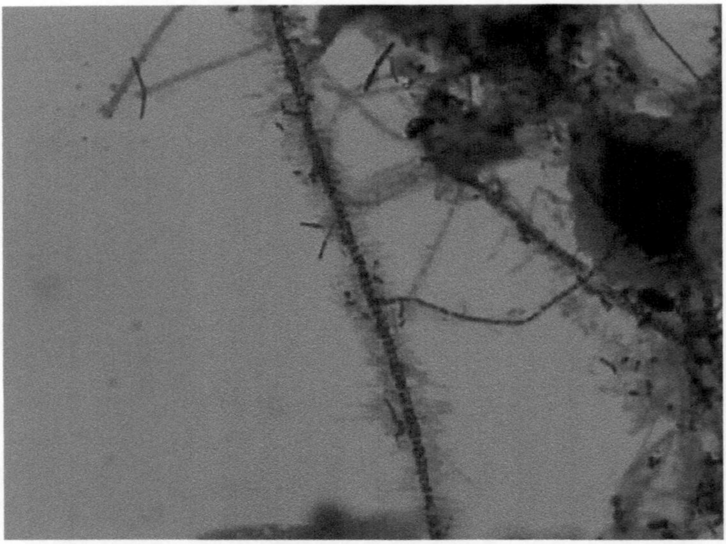

40. Type 0675_2
Gram variable type 0675 with a sheath
(Wet mount, 1000× phase contrast).

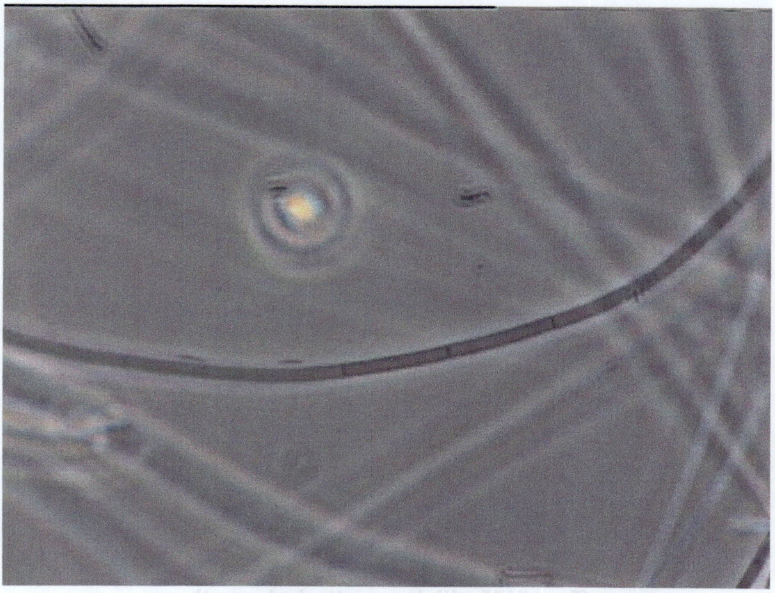

41. Type 0961_1
Type 0961 (Wet mount, 1000× phase contrast).

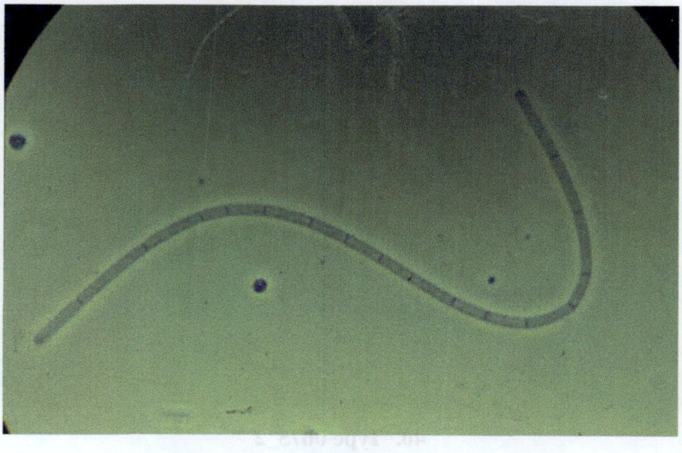

42. Type 0961_2
Type 0961 (Wet mount, 1000× phase contrast).

43. Type 1701_No growth
Type 1701 is occasionally seen
without attached growth
(Wet mount, 1000× phase contrast).

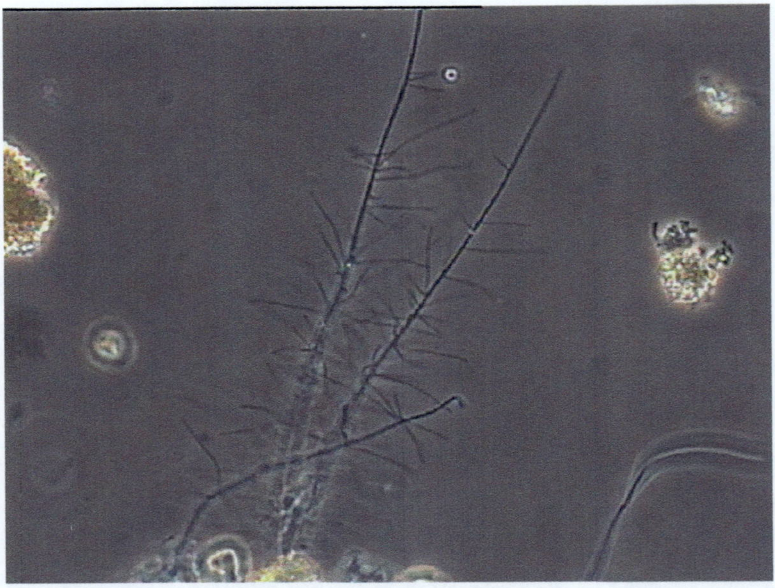

44. Type 1701_Attached growth
Type 1701 with heavy attached growth
(Wet mount, 1000× phase contrast).

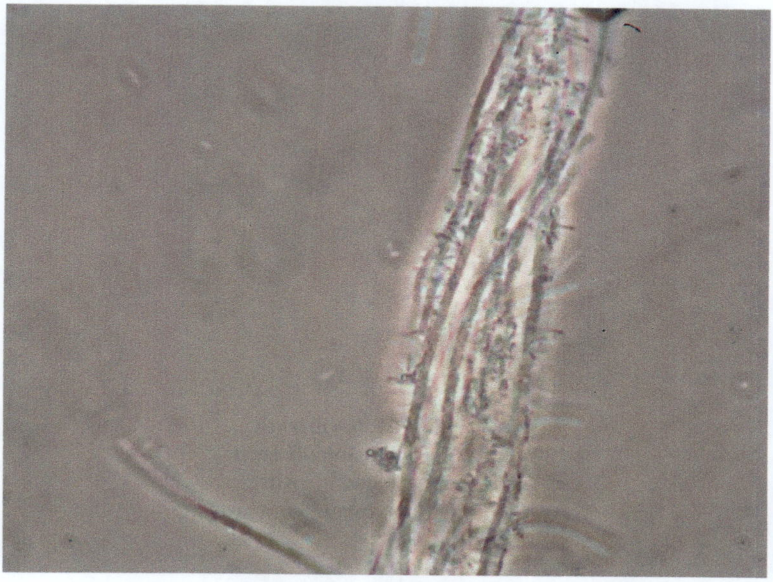

45. Type 1851_Bundle_1
Type 1851 bundle (Wet mount, 1000× phase contrast).

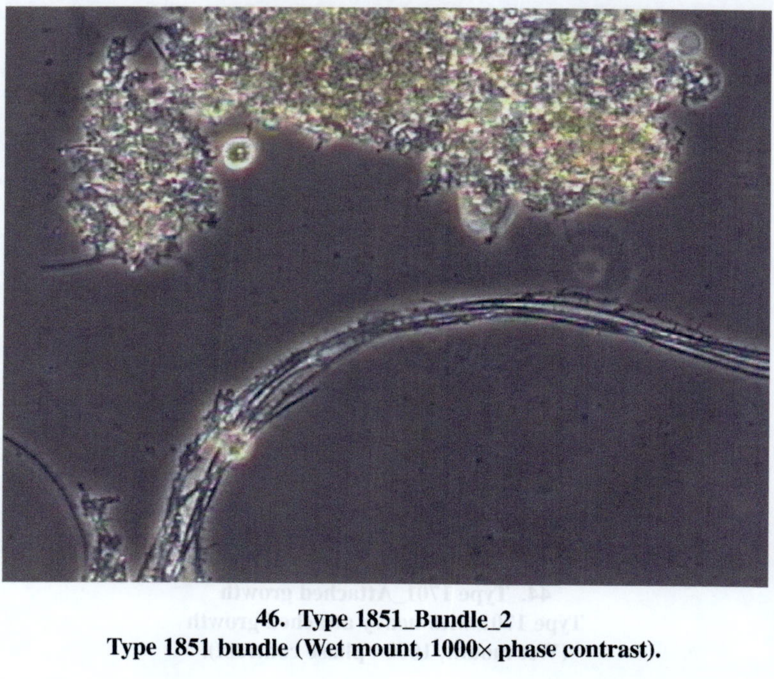

46. Type 1851_Bundle_2
Type 1851 bundle (Wet mount, 1000× phase contrast).

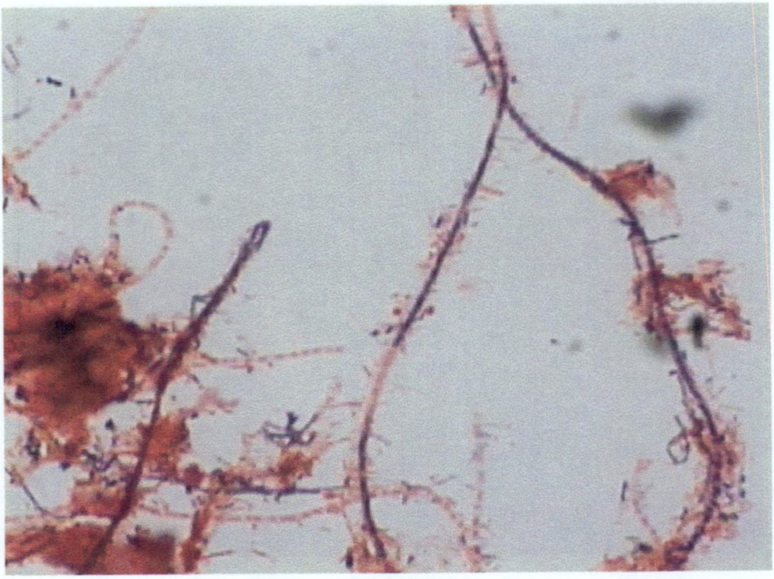

47. Type 1851_GramVar
Gram variable type 1851 (Gram stain, 1000 × brightfield).

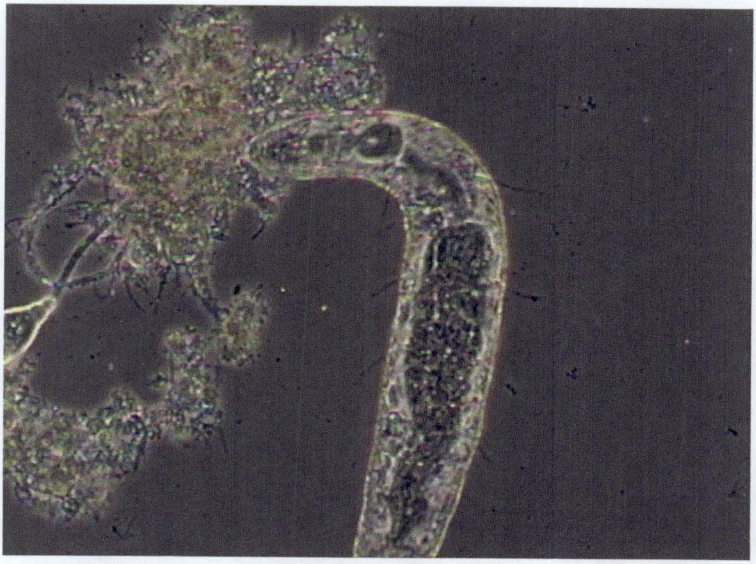

1. Bristle worm
Bristle worm (Wet mount, 400× phase contrast).

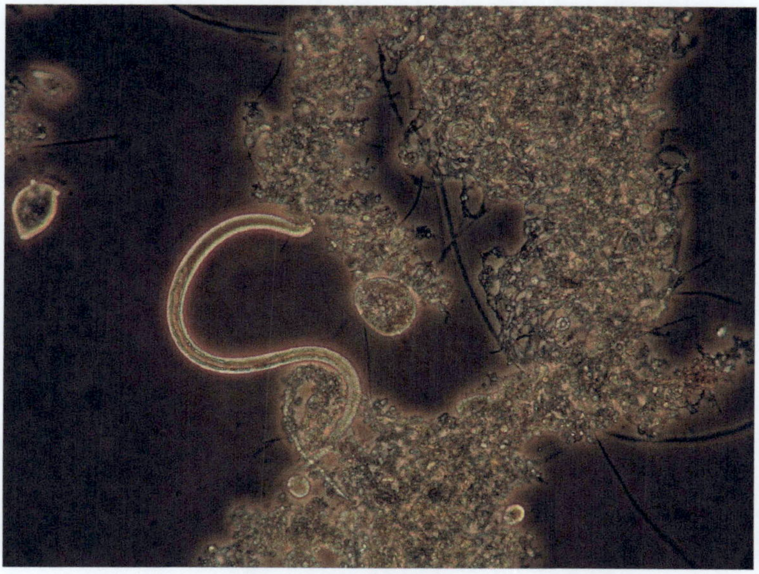

2. Nematode
Nematode (Wet mount, 400× phase contrast).

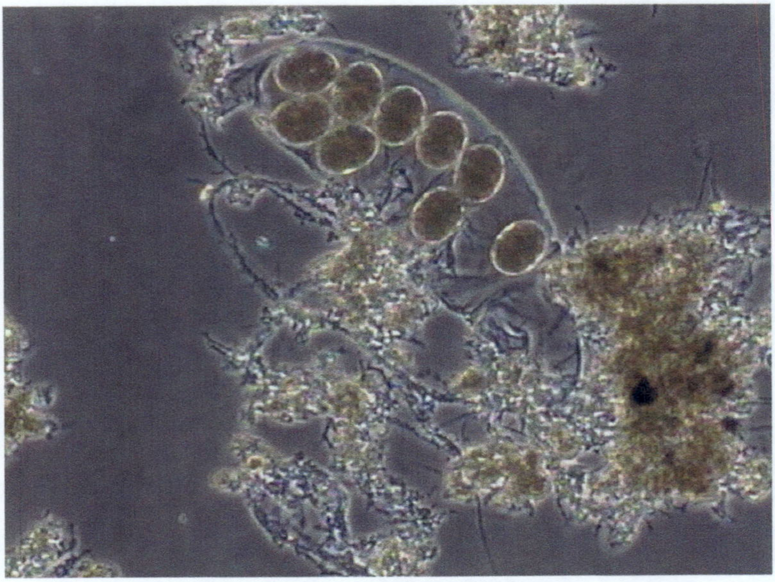

3. Pregnant Waterbear
Female waterbear with eggs (Wet mount, 1000× phase contrast).

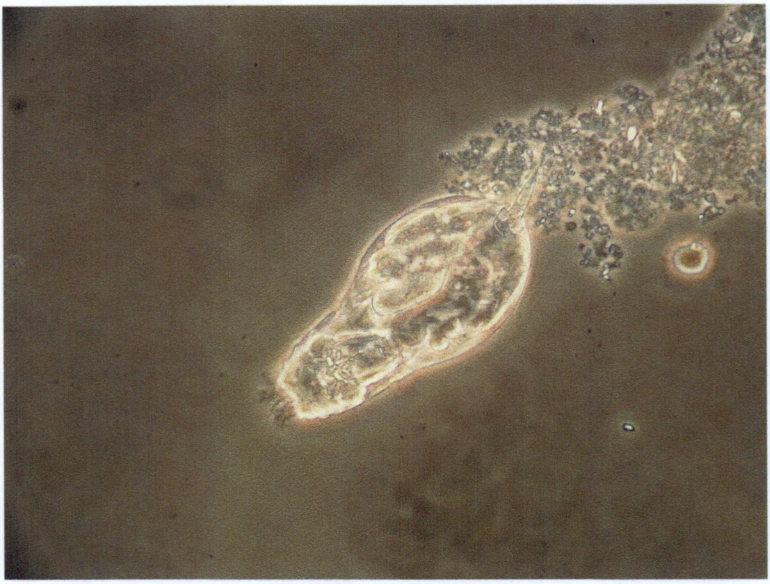

4. Rotifer_Brachionus
Rotifer, Brachionus sp. (Wet mount, 1000× phase contrast).

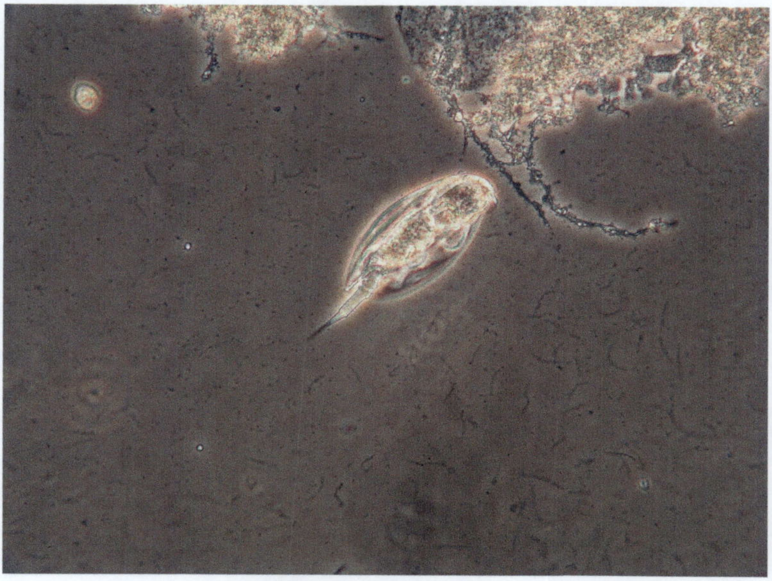

5. Rotifer_Lecane
Lecane sp. A shelled rotifer (Wet mount, 400× phase contrast).

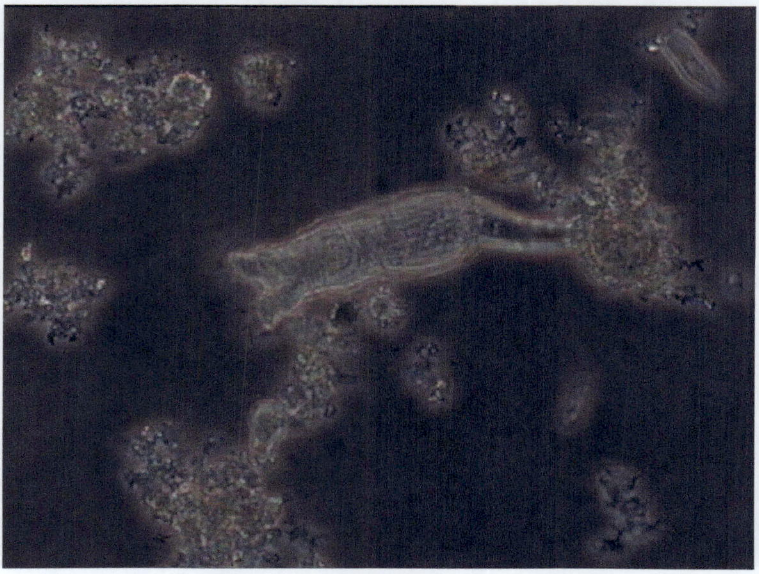

6. Rotifer_Rotaria_1
Rotifer, Rotaria sp. (Wet mount, 400× phase contrast).

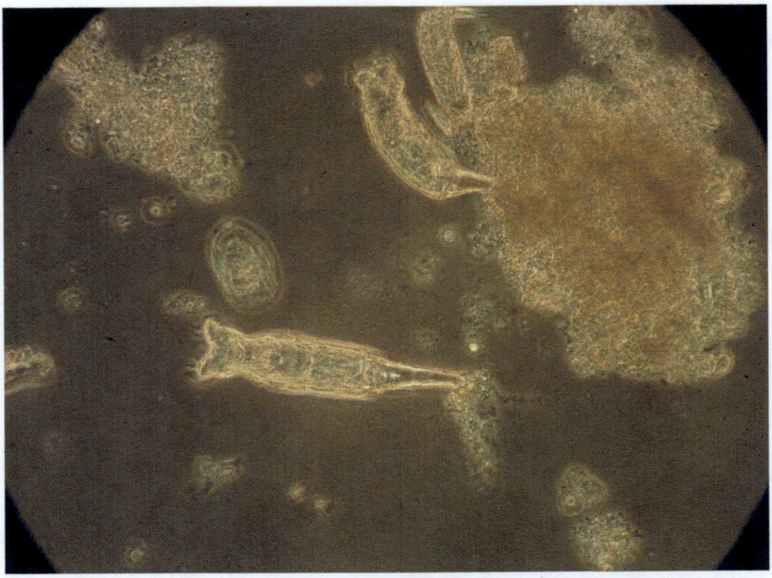

7. Rotifer_Rotaria_2
Rotifer, Rotaria sp. (Wet mount, 400× phase contrast).

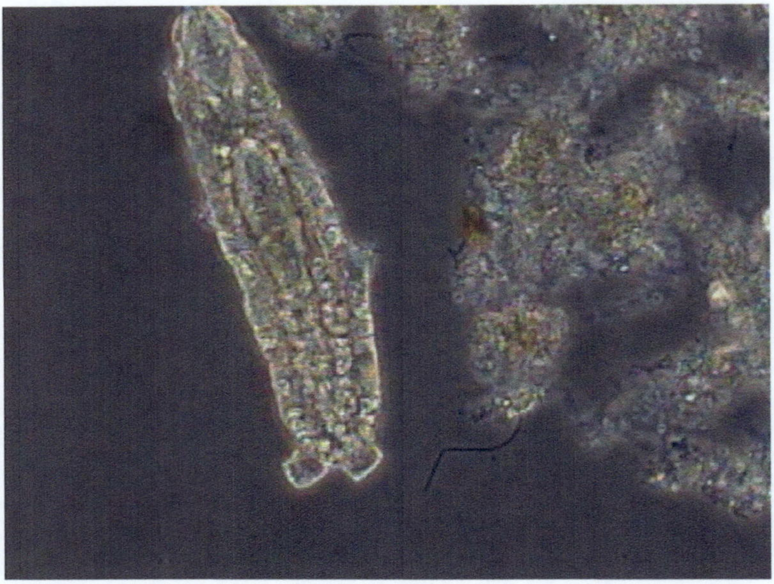

8. Waterbear
Tartigrade (waterbear) (Wet mount, 400× phase contrast).

1. Amoebae Arcella_1
Amoebae, Arcella (Wet mount, 1000× phase contrast).

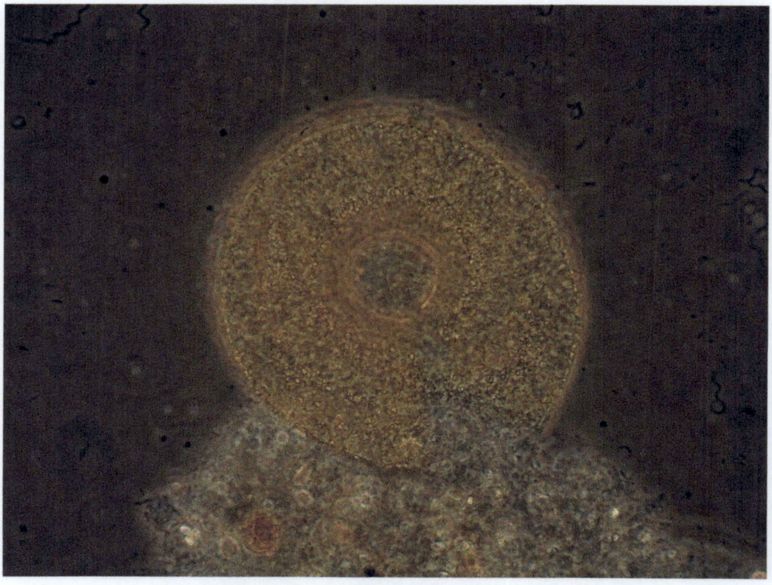

2. Amoebae Arcella_2
Amoebae, Arcella (Wet mount, 1000× phase contrast).

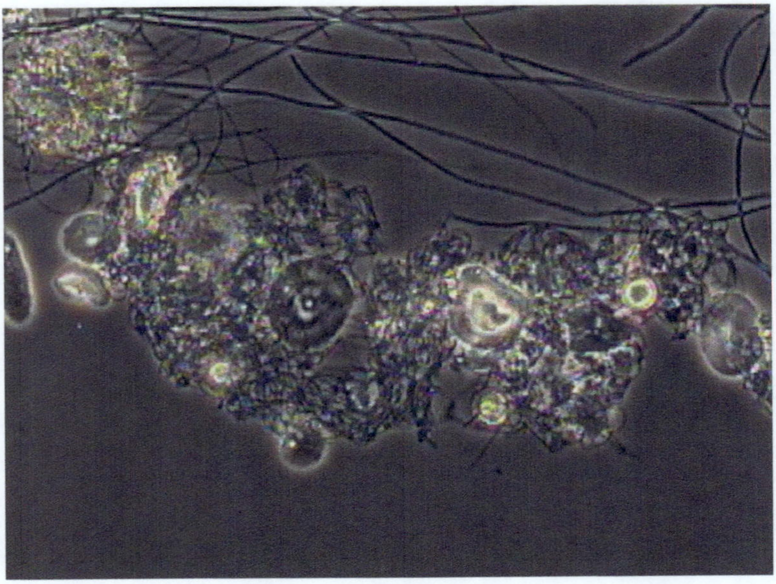

3. Amoebae Arcella_3
Amoebae, Arcella (Wet mount, 1000× phase contrast).

4. Amoebae Proteus_1
Amoeba, Proteus (Wet mount, 1000× phase contrast).

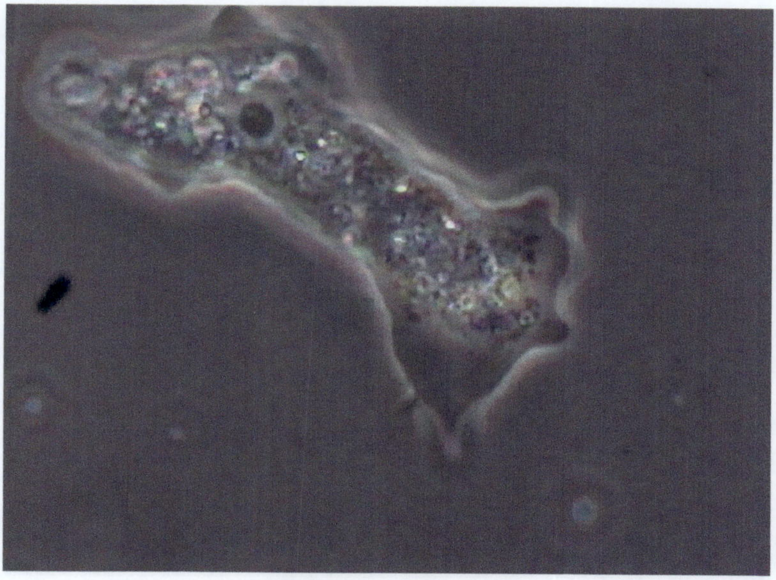

5. Amoebae Proteus_2
Amoebae, Proteus (Wet mount, 1000× phase contrast).

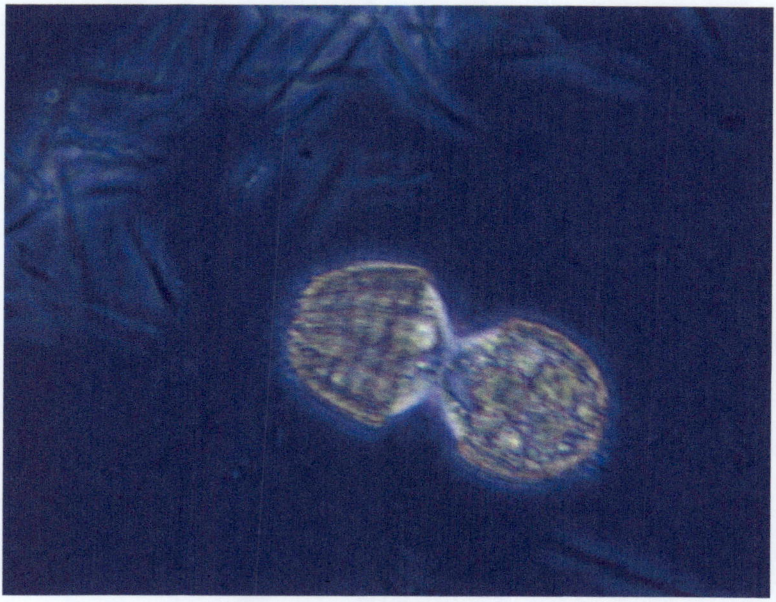

6. Coleps
Coleps (1000× phase contrast with a blue filter).

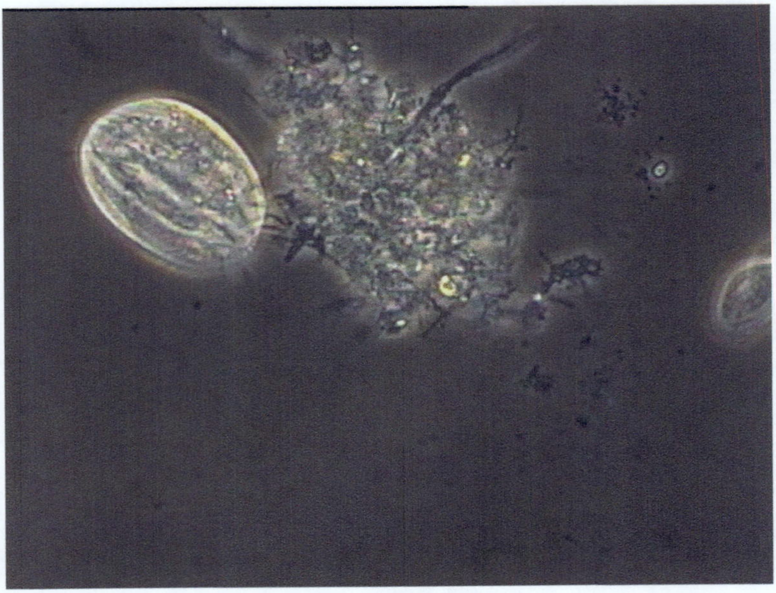

7. Aspidisca_1
Aspidisca, a crawling ciliate (Wet mount, 1000× phase contrast).

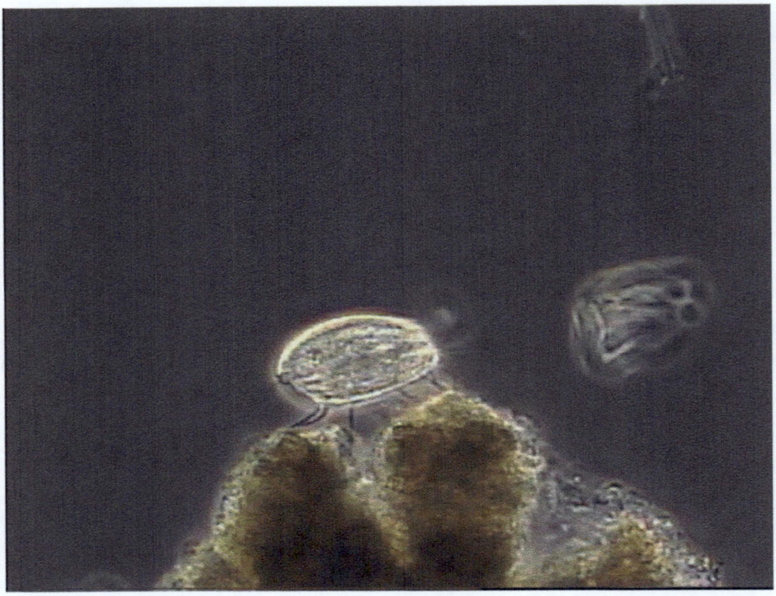

8. Aspidisca_2
Aspidisca, a crawling ciliate (Wet mount, 1000× phase contrast).

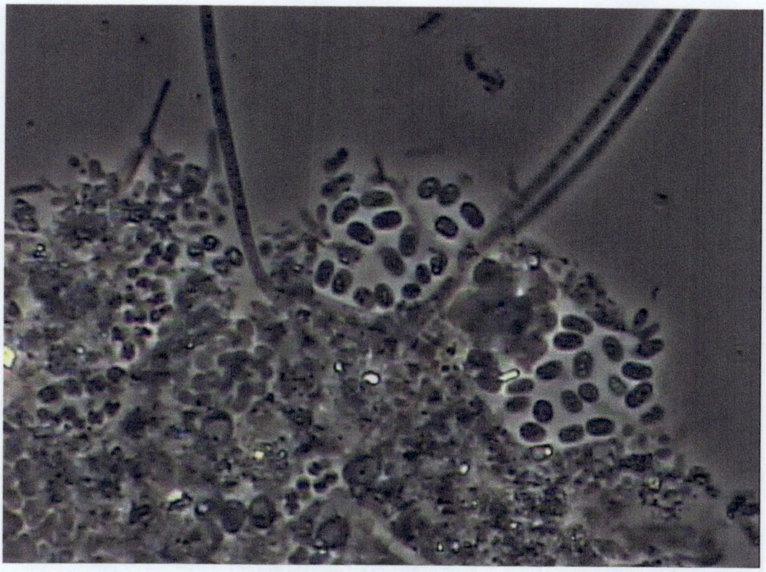

9. Spongomonas
Colony of flagellates (Wet mount, 1000× phase contrast).

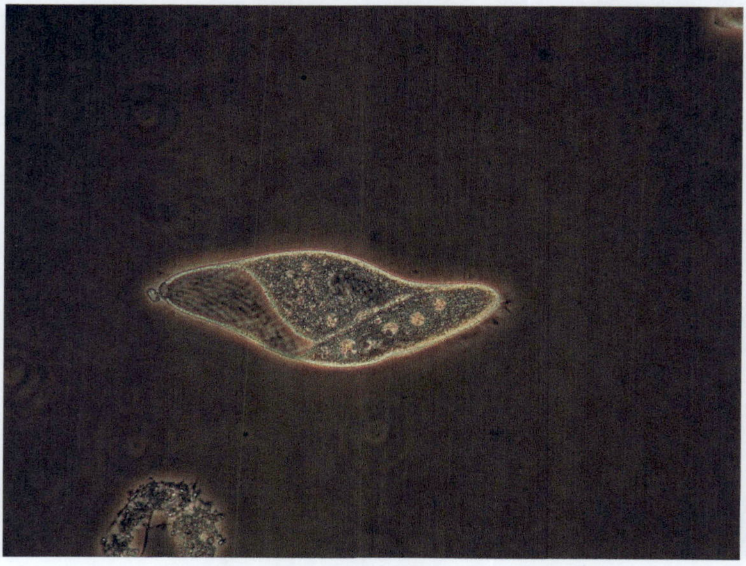

10. Paramecium
Paramecium (Wet mount, 1000× phase contrast).

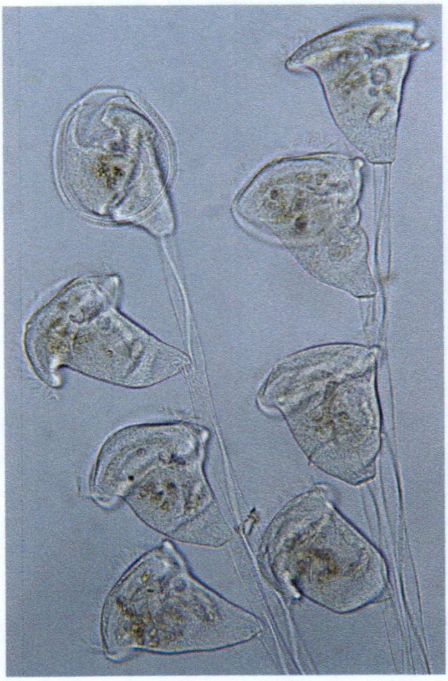

11. Carchesium_1
Carchesium, a colonial stalked ciliate (Wet mount, 1000× phase contrast).

12. Carchesium_2
Carchesium, a colonial stalked ciliate (Wet mount, 1000× phase contrast).

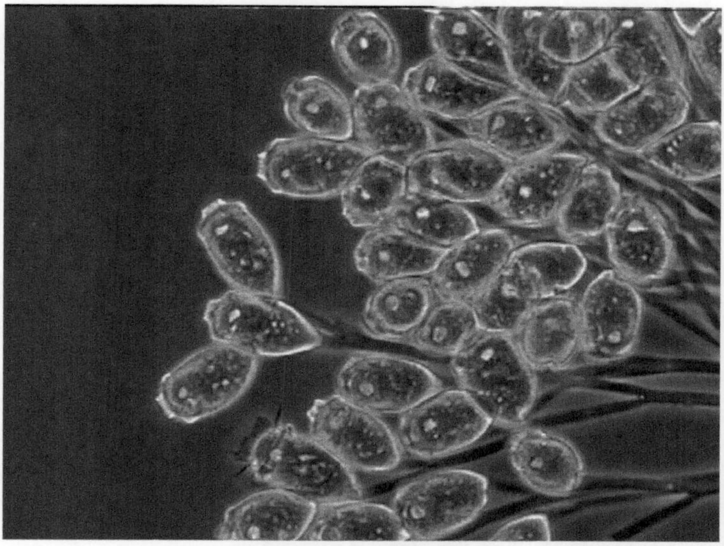

13. Epistylis_1
Epistylis, a colonial stalked ciliate
(Wet mount, 1000× phase contrast).

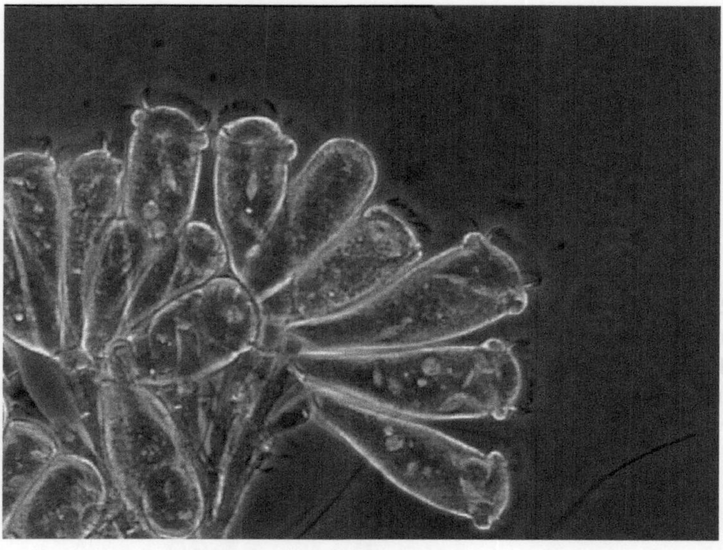

14. Epistylis_2
Epistylis, a colonial stalked ciliate
(Wet mount, 1000× phase contrast).

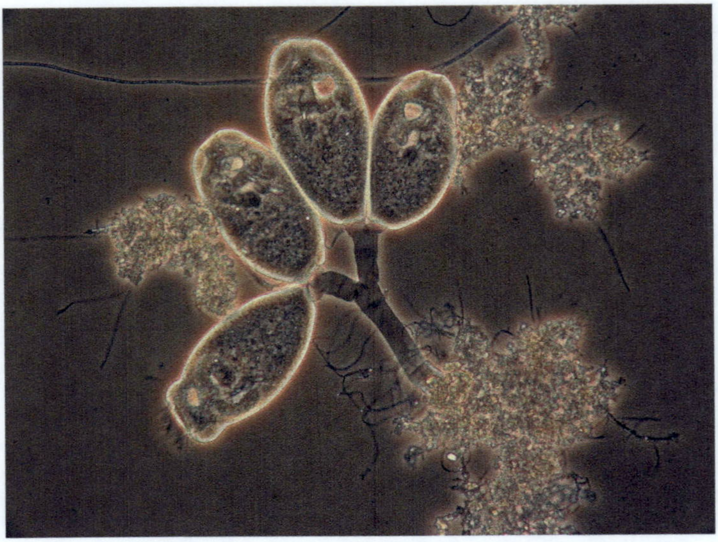

15. Opercularia_1
Opercularia, a colonial stalked ciliate (Wet mount, 1000× phase contrast).

16. Opercularia_2
Opercularia, a colonial stalked ciliate (Wet mount, 1000× phase contrast).

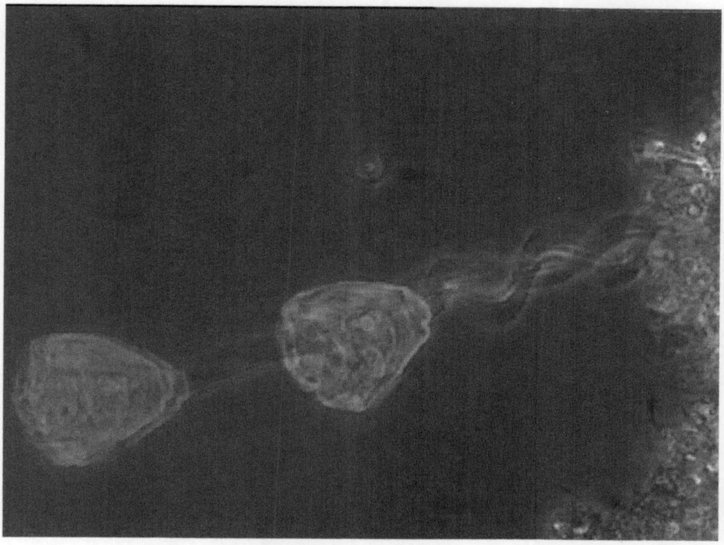

17. Vorticella_1
Vorticella. This single stalked ciliate recoils when disturbed (Wet mount, 1000× phase contrast).

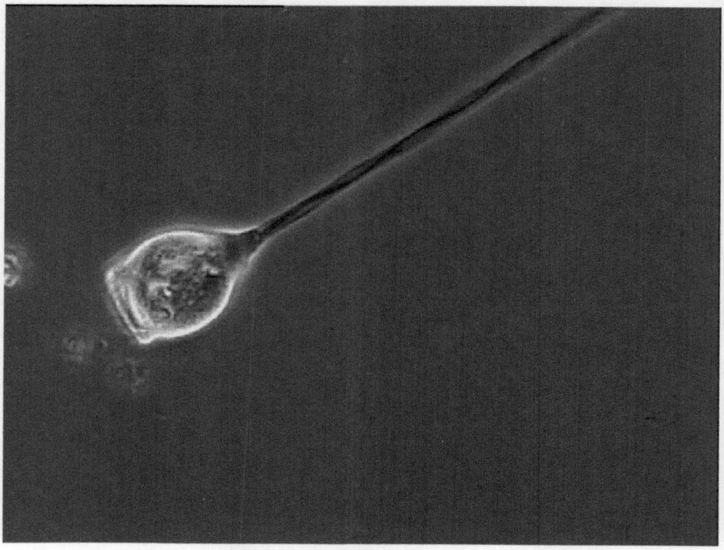

18. Vorticella_2
Vorticella. The myoneme or muscle fiber is clearly seen within the stalk (Wet mount, 1000× phase contrast).

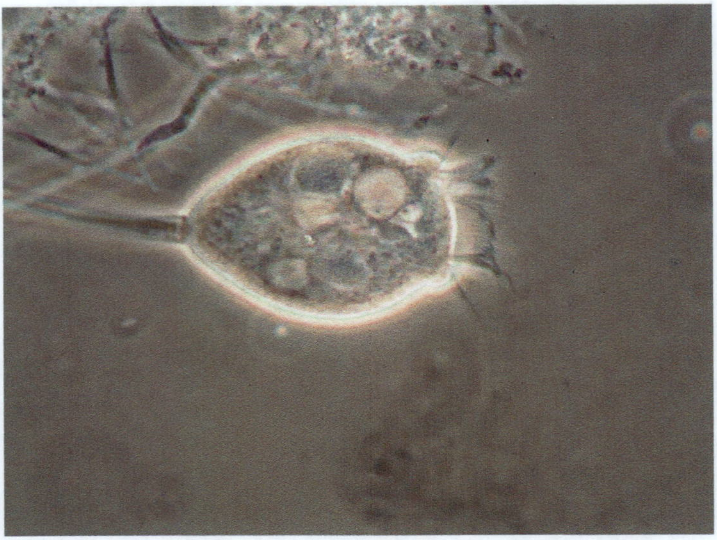

19. Vorticella_3
Vorticella. (Wet mount, 1000× phase contrast).

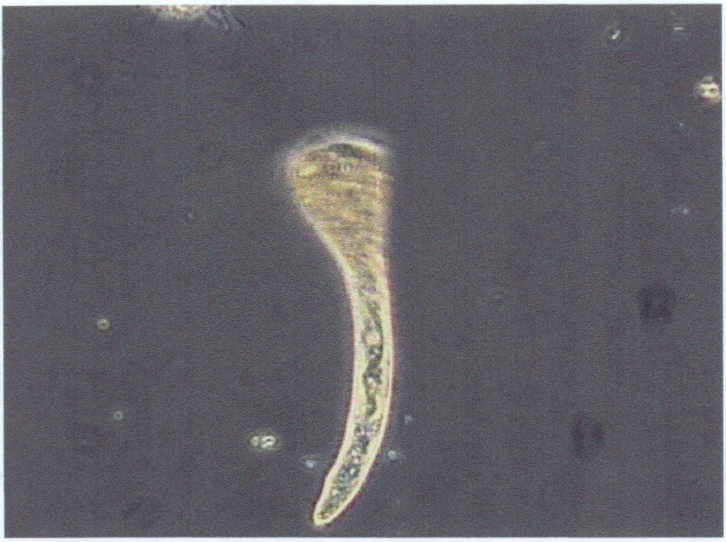

20. Stentor_1
Stentor (Wet mount, 1000× phase contrast).

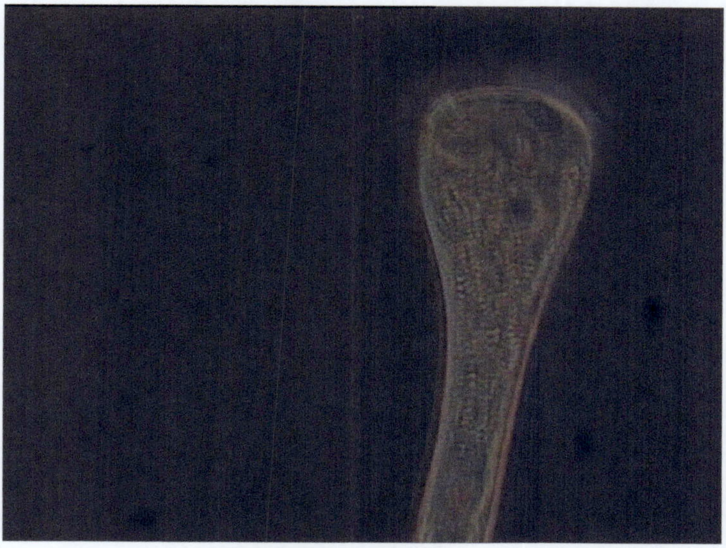

21. Stentor_2
Stentor (Wet mount, 1000× phase contrast).

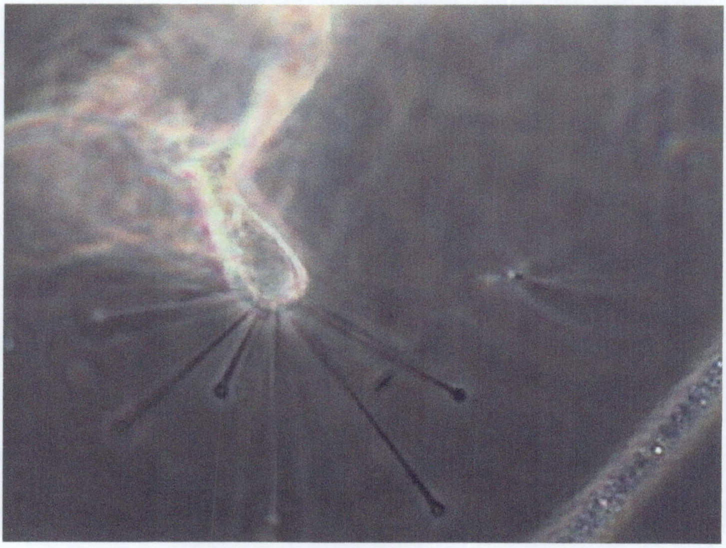

22. Suctorian_1
Suctorian tentacles (Wet mount, 1000× phase contrast).

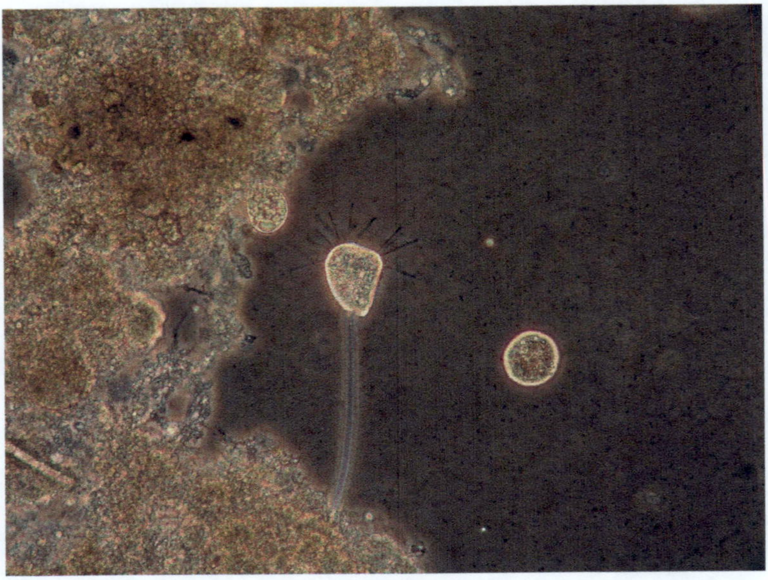

23. Suctorian_2
Suctorian (Wet mount, 1000× phase contrast).

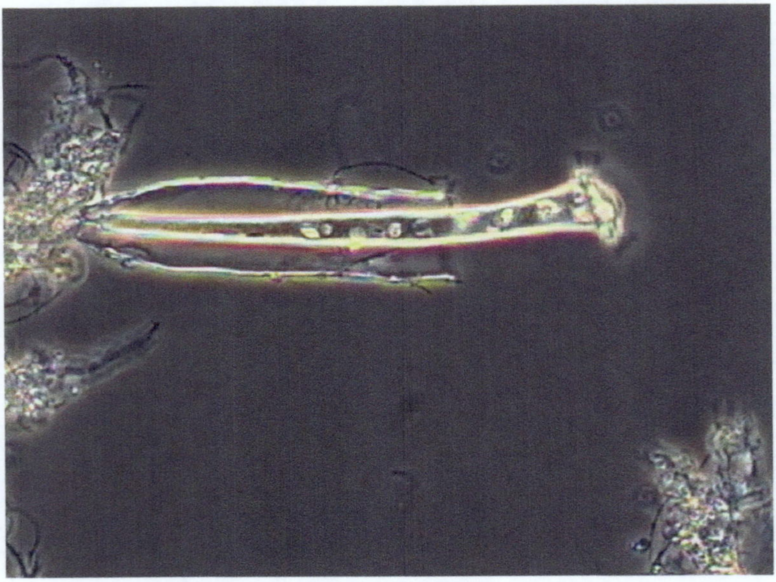

24. Vaginocola
Vaginocola (Wet mount, 1000× phase contrast).

Glossary

Absorption: The taking in, incorporation, or reception of soluble nutrients into the cell membrane.

Activated sludge: Sludge particles produced in raw or settled wastewater (primary effluent) by the growth of organisms (including zoogleal bacteria) in aeration tanks in the presence of dissolved oxygen. The term *activated* comes from the fact that the particles teem with bacteria, fungi, and protozoa. Activated sludge is different from primary sludge in that the sludge particles contain many living organisms, which can feed on the incoming wastewater.

Adsorption: The property of a solid substance to adhere and hold to the surface a cell.

Aerobic: A condition in which free or dissolved oxygen is present in the aquatic environment.

Anaerobic: A condition in which free or molecular oxygen is *not* present in the aquatic environment.

Anoxic: A condition in which only molecular oxygen is present.

Autotroph: An organism that uses carbon dioxide as its principal carbon source.

Bacillus: Any rod-shaped bacterium. When written as a genus, refers to rod-shaped, endospore-forming, facultative anaerobic, gram-positive bacteria.

Bacterium: A single-celled microorganism with a prokaryotic cell.

Biochemical oxygen demand (BOD): A process that measures the rate of oxygen use under controlled conditions of time and temperature. Used to estimate the organic loading to the wastewater treatment plant.

BOD: See biochemical oxygen demand.

Brightfield microscopy: The use of visible light for illumination; specimens are viewed against a white background.

Bulking sludge: Clouds of billowing sludge that occur throughout secondary clarifiers and sludge thickeners when the sludge will not settle properly.

Catalyst: A substance that affects the rate of a chemical reaction. It usually increases the rate, but isn't changed in the reaction.

Chemical oxygen demand (COD): A measure of the oxygen-consuming capacity of organic matter present in wastewater.

Clarifier (settling tank, sedimentation basin): A tank or basin in which wastewater is held for a period of time, during which the heavier solids settle to the bottom and the lighter materials float to the water surface.

Coccobacillus: A bacterium that is an oval-shaped rod.

Coccus: A bacterium of round, spherical, or ovoid form.

COD: See chemical oxygen demand.

Cytostome: The mouth-like opening in some protozoa.

Darkfield microscopy: The use of a device to scatter light from the illuminator so the specimen appears white against a black background.

Detention time: The time required to fill a tank at a given flow or the theoretical time that the microorganisms remain in the aeration basin.

Differential interference contrast (DIC) microscopy: The use of a microscope that provides a three-dimensional image.

Diplobacillus: A rod that divides and remains attached in a pair.

Diplococcus: A coccus that divides and remains attached in a pair.

Effluent: Wastewater or other liquid—raw or partially or completely treated—that flows from a basin, treatment process, or treatment plant.

Electron microscopy: The process of using a flow of electrons instead of light to produce an image.

Endoenzyme: An enzyme that catalyzes reactions within the cell.

Enzyme: A protein that catalyzes chemical reactions in a living organism.

Exoenzyme: An enzyme that catalyzes reactions outside the cell.

Facultative: The ability to use either molecular oxygen or oxygen obtained from food materials such as sulfate or nitrate ions. Facultative bacteria can live under aerobic or anaerobic conditions.

Flagellum: A thin, whip-like appendage that arises from the surface of a cell and is used for cellular locomotion.

Floc-forming bacteria: Groups or clumps of bacteria and particles that have come together and formed a cluster.

Food to microorganism ratio: A measure of food provided to bacteria in an aeration tank.

Fungus: Any of numerous spore-bearing plants, including yeasts, molds, mildews, and mushrooms, that have no chlorophyll.

Gram stain: A differential stain that divides bacteria into two groups, gram-positive and gram-negative.

Heterotroph: An organism that requires an organic carbon source.

Hyphae: Long filaments of cells in fungi or *Actinomycetes*.

Metabolism: The sum of all the chemical reactions that occur in a living cell.

Metazoan: A multicellular animal organism in which the cells are differentiated and form tissues.

Mixed liquor: When the activated sludge in an aeration tank is mixed with primary effluent or the raw wastewater and return sludge, this mixture is referred to as mixed liquor as long as it is in the aeration tank.

Mixed liquor suspended solids (MLSS): Suspended solids in the mixed liquor of an aeration tank.

Mixed liquor volatile suspended solids (MLVSS): The organic or volatile suspended solids in the mixed liquor of an aeration tank.

MLSS: See mixed liquor suspended solids.

MLVSS: See mixed liquor volatile suspended solids.

Phase-contrast microscopy: The use of a compound light that allows examination of structures inside cells through a special condenser.
Protozoan: A unicellular microorganism that contains a nucleus.
Pseudopod: Extension of a cell that aids in locomotion and feeding.
Sarcinae: A group of eight bacteria that remain in a packet after dividing.
Slime layer: Viscous covering on some bacteria composed of a polysaccharide or polypeptide.
Sludge age: A measure of the length of time a particle of suspended solids has been retained in the activated sludge process.
Sludge (biosolids): The settleable solids separated from liquids during processing or the deposits of foreign materials on bottoms of streams or other bodies of water.
Smear: A thin film of material on a slide.
Spirillum: A spiral or corkscrew-shaped bacterium.
Spirochete: A corkscrew-shaped bacterium with an axial filament.
Staphylococcus: A broad sheet of cells.
Streptobacillus: A rod that remains attached in a chain after cell division.
Streptococcus: A coccus that remains attached in a chain after cell division.
Tetrad: Cocci in a group of four.
Vibrio: A curved or comma-shaped bacterium.

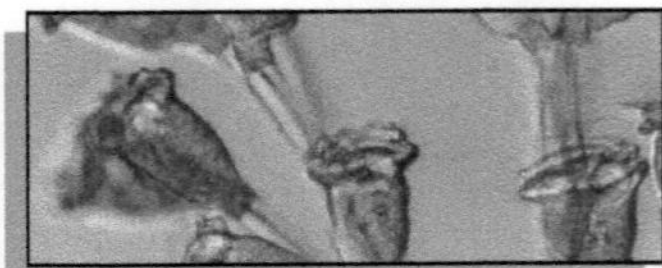

Bibliography

Lacy, J. 1988. *Actinomycetes as Biodeteriogens and Pollutants.* London: Academic Press. pp. 413–414.

Lemmer, H. 1985. The Ecology of Scum Causing Actinomycetes in Sewage Treatment Plants. *Jour. Sys. Appl. Micro.*

Tortora, G., Funke, B., and Case, C. 1982. *Microbiology: An Introduction.* Menlo Park, Calif.: The Benjamin/Cummings Publishing Co.

Waksman, S.A. 1981. *Actinomycetes, Vol. II: Classification, Identification and Description of Genera and Species.* Baltimore, Md.: Waverly Press.

Index

NOTE: f indicates figure; p indicates photo; t indicates table.

About the Author

Toni Glymph is a certified wastewater treatment plant operator with more than 12 years of hands-on wastewater treatment system experience. She has spent more than 22 years conducting microscopic evaluations and troubleshooting wastewater treatment system upsets. She has conducted more than 100 wastewater microbiology workshops including Wastewater Microbiology and Process Control, Filamentous Bacteria Identification and Control, Lagoon Treatment System Microbiology, Algae Identification in Wastewater Treatment Lagoons, Slime Bulking and Foaming, and The Use of Staining in Wastewater Microbiology Process Control. She continues to conduct an average of 12 workshops each year.

Toni is the recipient of the State of Wisconsin John C. Brogan Award for Outstanding Environmental Achievement for her work in helping operators achieve and maintain compliance of treatment system requirements. She has also received national praise for her work with wastewater treatment plant operators in training sessions and through one-on-one technical assistance. She is widely regarded as a top expert in her field and an excellent teacher, and has the ability to forge connections with her audiences. She is asked many times each year to teach wastewater treatment courses for the US Environmental Protection Agency (USEPA) and other state and national organizations. She has also

been invited to write a chapter in an international book on wastewater treatment.

Toni is also the recipient of the USEPA National Spark Plug Award for Operator Trainers, "a prize presented to the person or group that is at the forefront of the Operator Training Program and has made positive things happen to ensure the program's continuation." Russ Martin from USEPA headquarters states, "She dedicates her whole self to an issue and has given wastewater operators help in understanding the microbiology of what makes their plants work."

In addition to onsite assistance, the author conducts microscopic evaluations on wastewater samples she receives from treatment systems throughout the United States. She is also called on by wastewater treatment system design engineers and consultants for assistance in designing treatment system components that favor the growth of microorganisms best suited to treating wastewater.

Toni lives in Westport near Madison, Wisconsin. She has been married to her husband Rubin for 22 years. Toni helped raise two stepchildren, a son Rubin Jr., 26, and Toi, 25, and has two sons of her own, Lawrence, 20, and Daniel, 14.